Z. P.

Starting Points

Starting Points

for teaching mathematics in
middle and secondary schools

Colin Banwell
Ken Saunders
Dick Tahta

Oxford University Press 1972

Oxford University Press, Ely House, London W.1
GLASGOW NEW YORK TORONTO MELBOURNE WELLINGTON
CAPE TOWN IBADAN NAIROBI DAR ES SALAAM LUSAKA ADDIS ABABA
DELHI BOMBAY CALCUTTA MADRAS KARACHI LAHORE DACCA
KUALA LUMPUR SINGAPORE HONG KONG TOKYO

All photographs by Adrian Lack,
Oxford Illustrators Ltd

Photoset by BAS Printers Limited
Wallop, Hampshire
and printed in Great Britain at the
University Press, Oxford, by Vivian Ridler,
Printer to the University.

Preface

This book is a collection of suggestions for the teaching of mathematics. In the first part we assemble various comments and questions on ways of working in the classroom. In the middle section we give about forty suggestions for starting points, or situations from which mathematics can be made. Finally, in the last part, we list about twenty types of material that can be used in classrooms, with an indication of some of the mathematics that can arise. We have tried to be descriptive, in the sense that we have given an account—necessarily condensed—of work that has happened at some time. We have tried not to be prescriptive, but this will depend on the reader. The blank spaces which we have provided are also part of the text, to remind the reader that there are many options. We think that this is a book to be dipped into, not a comprehensive or consecutive account.

Most of the material has been developed from work in the Shelley and Queen Elizabeth's schools in Crediton, Devon, where two of the authors (C.B. and K.S.) were formerly heads of mathematics departments. We owe more than we can say to the many colleagues and children, at Crediton and elsewhere, who have taught us so much. In particular, we would like to acknowledge the great privilege it has been to work with such skilled and sensitive teachers as George Aplin, Danny Favis, Bill Maltby, Geoff Saltmarsh and Dorothy Watson.

We are grateful to the many children whose work we quote, explicitly or implicitly, throughout the book. They are too numerous to list but our appreciation of the importance of their contribution is in no way lessened by this brief acknowledgement. It is equally difficult to express our debts to the many people who have shared their ideas and experience with us. It would be impossible to trace the original source of much that we have found fruitful but we know how much we owe to the fertile imagination of others.

We wish to thank the authors of the privately published *Crediton Mathematical Reports*, nos. 1–5, for permission to use material from them; the Oxford University Press, for permission to quote from *Tales of the Hodja*; the United Feature Syndicate for our use of a *Peanuts* cartoon. Finally we would like to thank the publishers for their skilled and patient assistance.

C.B., K.S., D.T.

Contents

I Ways of working

II Starting points

III Materials for mathematics

IV Appendix

I Ways of working

I Ways of working

introduction

There have been some bewildering developments in the teaching of mathematics over the last two decades. Before these changes, a mathematics teacher could be reasonably certain that, for the rest of his career, he would be expected to instruct his pupils in a fixed collection of ideas and techniques, which were simplified and systematised in standard text books. He would not need to know very much about the development of mathematics since the seventeenth century. Moreover the methods he would be expected to use were those by which he himself was taught. The classroom in which he taught appeared the same as the one in which he learned and there would be little difference if he moved from school to school. There was no shortage of wise experience to guide him when he started teaching, and a settled educational system provided objectives to work for, and criteria for success and failure.

There were, of course, many snags. Mathematics seemed an infernally difficult subject to teach and most of the adult population seemed to be dramatically scarred by their experience of having to learn it. 'I was never any good at mathematics' was—and still is—a common conversational opening on meeting a mathematics teacher, to be followed by traumatic memories of some childhood experience. Even the most intelligent did not cope very well; half the girls selected for grammar schools never even sat for public examinations in mathematics. As for the vast majority of children, misconceptions about the nature of abstract thinking led dangerously close to an abdication from mathematics in favour of a diluted course of practical science and some so-called civic or domestic arithmetic.

This is not the place to review the changes that have taken place since the second world war. The struggle to change the syllabus, to extend the range of mathematics that is to be learned in schools, has sometimes tended to obscure the changes in ways of working in the classroom. And such changes have hardly begun in the majority of schools. But it is clear that the teacher of mathematics will have to meet and develop a changing situation in the foreseeable future. For the young teacher entering the profession in the future, there is no clear-cut expectation of what he can and should do. There is a bewildering variety of text books following different courses and leading to different public examinations.

His subject seems to be growing in many directions and he may have to work at bits of mathematics being developed in his own time. His classroom may or may not look the same as the one in which he was taught, but in any case he has many choices of ways he can change it. His students may soon be challenging his authority. If he changes job he may find himself in a totally different situation, and the system he is working in faces reorganisation for some time. The prospect is alarming but could be fruitful.

The Hodja was walking home with a fine piece of liver when he met a friend.

'How are you going to cook that liver?' asked the friend.

'The usual way,' said the Hodja.

'That way it has no taste,' said the other. 'I have a very special way of preparing a very tasty meal with liver. Listen and I'll explain.'

'I am bound to forget it, if you tell me,' said the Hodja. 'Write it down on a piece of paper.'

The friend wrote out the instructions, and gave them to the Hodja, who continued on his way home. Before he arrived at his door, however, a large crow swooped down, seized the liver in its claws, and flew high up into the sky with it.

'It won't do you any good, you rogue!' shouted the Hodja triumphantly waving the piece of paper. 'I've got the recipe here!'

This, and other quotations in this section, come from *The Tales of the Hodja*, retold by Charles Downing and published by Oxford University Press. *Hodja* was an honorary title for a scholar, in particular one learned in the Koran and religious law. Stories about the folk-hero, Nasreddin Hodja, are said to have been used by the Sufi masters as a vehicle for their teaching.

Where to start? We have not got a recipe and we hope the reader has not got one either. Despite a popular view, mathematics itself is not a subject in which everything grows out of something that has gone before in a single thread reaching into the past. Nor can the learning of it always be linear, even though some teaching tries to be. In the same spirit we do not intend this book to be read from beginning to end as some gradually unfolding guide for the teaching of mathematics. We have assembled some starting points and have tried to indicate a few of the many roads that can be taken. There are no finishing posts.

an investigation

Where to start? We begin with an account of the way in which a class of children developed a simple mathematical situation. . . .

This figure was drawn on the blackboard. A boy was asked to draw in a diagonal and the class, a third form, was asked how many squares of the rectangle it passed through. There was some disagreement about the count. From the discussion amongst the class, this appeared to stem from different interpretations of what had been asked. The question was repeated and there was an agreed answer. The class was now asked to investigate this situation in general. The children worked actively in groups, making statements on the blackboard from time to time. After the first few minutes the teacher played no further public part in the lesson but sat and listened when anyone had something they wished to say.

It is sometimes said that mathematics is essentially problem-solving. This can be misleading if by this is meant the getting of answers to problems which are well-defined and understood. At the start of the mathematical activity a problem is rarely well-defined. Preliminary mental 'doodling' begins to clarify the outlines and very often the solution consists of a re-definition or a refinement of the problem so that it fits into the scheme developed during the doodling. This fitting and refitting is rarely mentioned in mathematical literature today, though some of the great classic papers are full of subjective accounts of the mental paths which led to the final result. The importance of mathematics in education lies in process rather than product. An actual formula for the number of squares the diagonal of a rectangle passes through is not, after all, an important thing to possess. The process of getting a formula is, and this process was ably and naturally described by the children on this occasion. The following extracts are taken from their writing about their investigation of the problem. The reader is urged to consider the problem himself before reading further.

extracts from children's work

—First, my rectangle was 7×5. The number of squares that the diagonal passed through was 11. My next number was 4×3. The number of squares was 6. So, so far it looked like if the two numbers were added together and 1 taken from them, this was the answer. I then tried again just to make sure, $9 \times 6 = 12$. My theorem was wrong. But I wasn't quite sure how many squares there was, so in the middle of the 9×6 I made a 3×2 and this made certain there was only 11 squares in which the diagonal passed through, so my theorem was wrong.

—To try and find a way to determine the number of squares in a rectangle I first started with similar rectangles that got bigger and bigger. I started with a rectangle 2×3 squares and then went to 6×9, 10×15, 14×21, 18×27 and 22×33 . . . then I made this table . . . (which) does not count any point as a square because the diagonal goes through a point for every four squares. Then I made another table but this time counting every point as a square . . . now these numbers looked like getting somewhere. But this was still not the right numbers to make a rule with because I used points but as you see these numbers go up in tens.

—This is now where the 2×2 square will be included: from it can be seen the fact that it passes through two squares. Coming back to the idea of adding the two numbers involved in the size and subtracting one, then the following statement would be true: $2+2-1=2$. But that is not true! Before all hope was given up the other numbers were tried, with the following results . . . it was seen then that it was $2+2$ (4) (6) that didn't work: 'two+an even number'. So we looked at these again. . . .

—To find the number of squares the diagonal passes through add the two numbers of the rectangle and subtract one. This only works if you count where the diagonal passes through an actual corner of the square.

extracts from children's work
—continued

—To find the number of squares it passes through multiply the number of squares on the length and breadth together and take away the number of 2×2 squares there are in the rectangle.

Another way is to count the number of squares around the inside of the rectangle and divide them by two.

—We then looked for a connection in numbers of the table. Finally the idea that by adding the two dimensions together, subtract one and you have the number of squares cut through by the diagonal. This works for some but not all . . . that idea did not work for 2×4 or 2×6. It was found that in each of these two the diagonal went through a point where four squares met.

—(i) dimension 1 + dimension 2 − 1; (ii) . . . same as the square has down the side; (iii) . . . nearly all the odd numbered dimensions seem to apply to the dim1 + dim2 − 1 formula except 5×15, 9×45, etc.; (iv) . . . nearly all the even-numbered dimensions seem to apply . . . except 4×6, 8×12, 16×24, etc. . . . also there are rectangles with dimension of $1 \times ?$ . . . all these have the same number of squares that the diagonal pass through as their length.

—If you add up the two dimensions of the rectangle and subtract 1 that will give you the number of squares which the diagonal passes through. . . . If you add up the number of squares around one half of the diagonal you will also get the answer to the number of squares which the diagonal passes through.

—Then I thought of something. The 7×5 could not have a smaller square inside it but the 9×6 could. So therefore my theorem worked but only for those numbers which could not be divided up in a smaller square. . . .

Problem

What number of squares does a diagonal pass through in any size rectangle divided into squares?

This is how we set about the problem. —
By drawing several size rectangles and comparing them a table was worked out. Underneath are the no. of squares passed through.

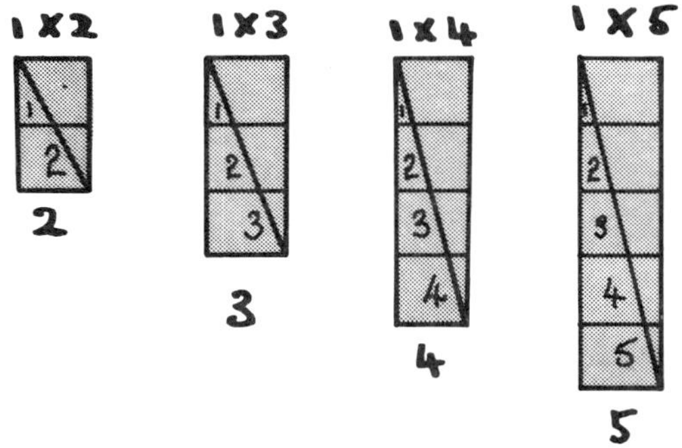

If you take the size of the ~~square~~ rectangle
eg. 1X4 add 1 to 4 and subtract one.
ie. 1 + 4 - ~~4~~ 1 you get the answer to the problem set.

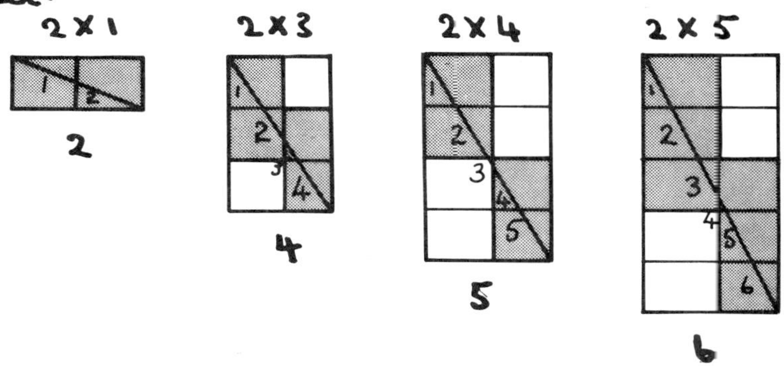

Again take the size of the ~~square~~ rectangle
eg. 2X5.
Add 2 to 5 and subtract 1
ie. 2+5-1 again you get the answer 6.

So these were looked into again:

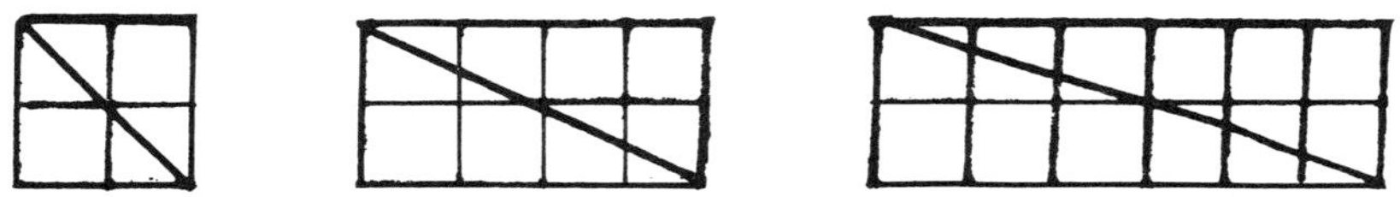

and it was seen that in each on of these the diagonal went through a point where four squares met:

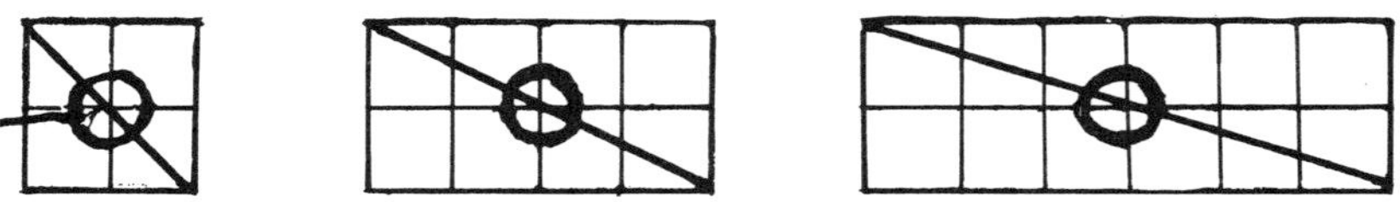

A slight problem here is to decide whether this goes through one, two or no squares!
So a reasonably safe statement to make at the moment is:
Add the 2 dimensions together and subtract 1, except when both dimensions are even.
The next move would then be to keep one dimension at 3. Then this would be the <u>following table</u>:

3 × 1	goes	through	3	squares,	with	0	meeting	pts.
3 × 2	"	"	4	"	"	0	"	"
3 × 3	"	"	3	"	"	2	"	"
3 × 4	"	"	6	"	"	0	"	"
3 × 5	"	"	7	"	"	0	"	"
3 × 6	"	"	8	"	"	2	"	"
3 × 7	"	"	8	"	"	1	"	"

3X1

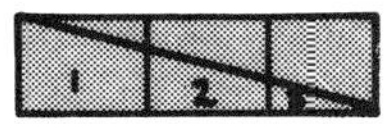

3

3X2

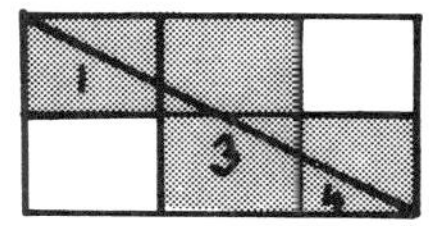

4

3X4

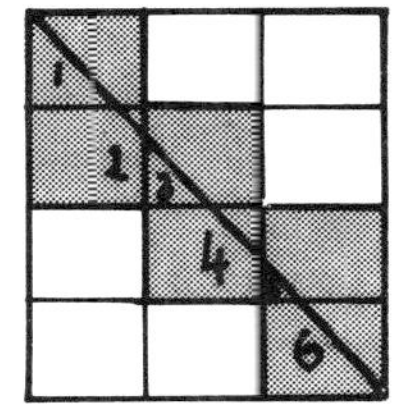

6

3X5

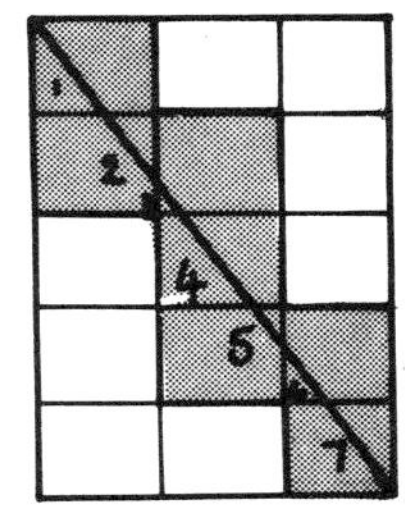

7

Take a rectangle.
3X2.
Add 3 to 2 and subtract 1
This time you get the answer 4

The tables run as follows.

2x1	2
2x2	–
2x3	4
2x4	5
2x5	6
2x6	7
2x7	8
2x8	9

3x1	3
3x2	4
3x3	–
3x4	6
3x5	7
3x6	8
3x7	9
3x8	10

4x1	4
4x2	5
4x3	6
4x4	–
4x5	8
4x6	9
4x7	10
4x8	11

5x1	5
5x2	6
5x3	7
5x4	8
5x5	–
5x6	10
5x7	11
5x8	12

You notice the numbers follow each other
ie. 4, 5, 6, 7, 8, 9,
If we were using squares this would also fit into the series of numbers.

There is always a series running the other way.
As below.

n = any number

2Xn		3Xn		4Xn		5Xn
2	←+1→	3	←+1→	4	←+1→	5
(3)	←+1→	4	←+1→	5	←+1→	6

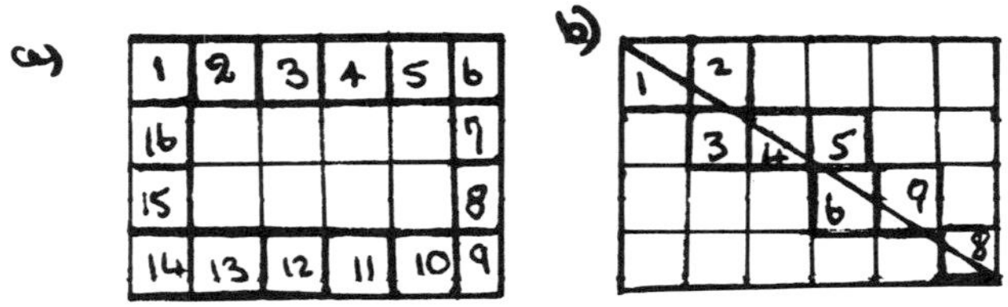

$a = 16$ sqs $b = 8$ sqs $\therefore a = 2b$.

On the squared paper given several rectangles were drawn and the diagonals put in, the number of squares that the diagonal passed through are then noted. A conclusion was drawn up after several of these had been thought about. If you take a ▭ 4×5 and draw the diagonal in and then count the squares there are 8. To find the number of squares through which the diagonal passes add 4 and 5 together and take away 1. Similarly if you had a ▭ with sides x and n the number of squares through which the diagonal passed would be $x+n-1$. This does not work in an even sided figure unless the intersection is counted as one. If the ▭ has sides of even numbers the diagonal will always intersect the number of intersections can be found by dividing the length of the smaller side into the length of the longer side. Another way of finding the number of squares the diagonal passes through is to count the number of square around the inside of the ▭ and divide them by two (see fig)

Problem: The number of squares which the diagonal of a rectangle pass through

First form a table.

Dimensions	Squares which diagonal pass through
2 x 3	4
3 x 4	6
4 x 5	8
5 x 6	10
6 x 7	12

Then try others as well

3 x 6	8
5 x 15	19.

Then find a similarity.

In this the addition of the dimensions minus one is equal to the number of squares which the diagonal pass through.

Let dimensions = d_1 and d_2

and the squares = n

$\therefore \quad n = d_1 + d_2 - 1$

Also whilst studying this other things were found.

1 If the two dimensions will divide into each other (eg 3, 15, 7, 21,) then at every 1 square one way and (with 3, 15) 5 the other way and intersection will be crossed.

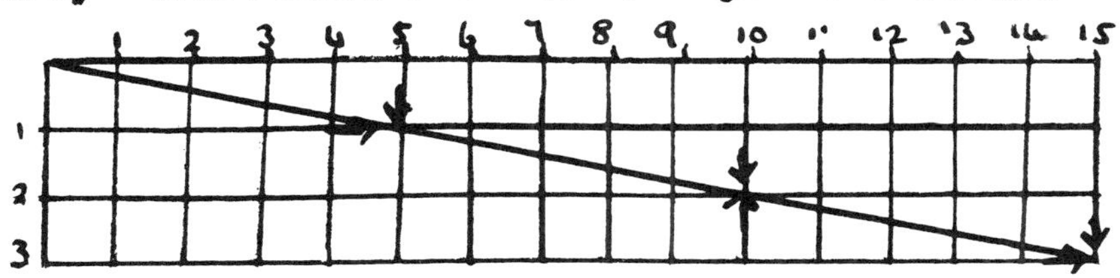

If you draw a retangle half way between squares. (see diagram below) so that they come into a whole numbers. then add the no. of square together it will give you how many squares it goes through.

5 by 3

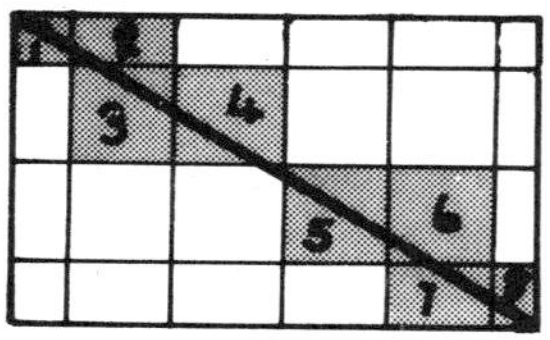

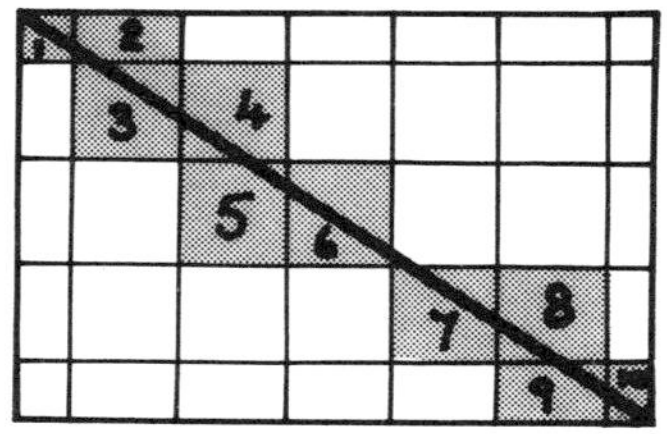

6 by 4

If you have a retangle 4 by 5 and multiply the two sides together, 4 x 5 = 20 .

Then find out how many 2x2 squares there are in the retangle. You can do this by taking one from 4 and 5 = 3 and 4. then multiply these two together which equals 12. Take 12 from 20 which Leaves 8.

If you have a retangle by $3\frac{1}{2}$ by $4\frac{1}{2}$ (or any other retangle with 2($\frac{1}{2}$)'s.). Then add them together which equals 8 . this is how many squares it goes through.

3 An intersection will be crossed at at least one point on the rectangle if it is a square or if the numbers will divide into each other or if both dimensions are even.

eg

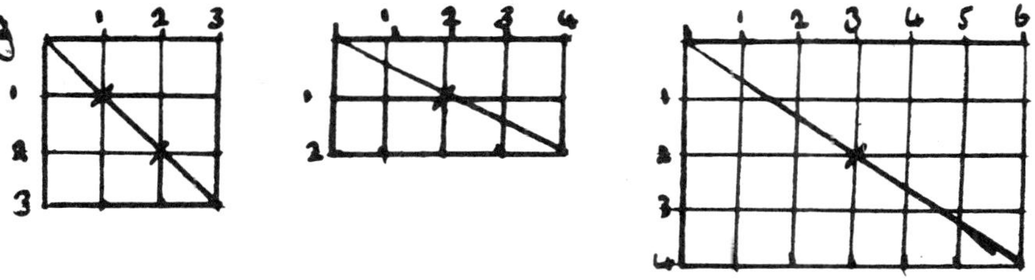

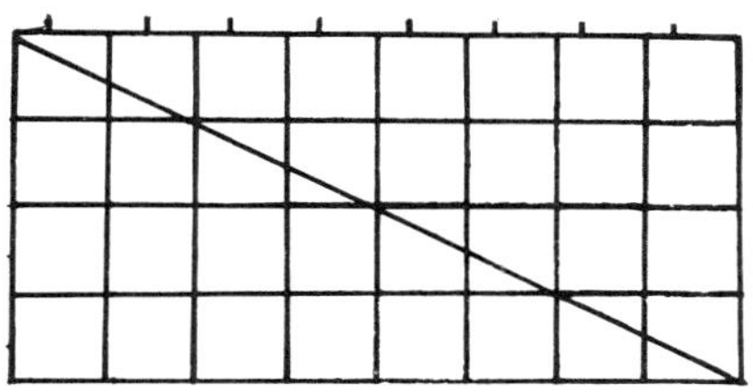

8 X 4

It is noticed that the line passes through 11 of the squares. This is only correct if the points which the line travels through are also counted as separate squares.

<u>Some Examples.</u>

2X6

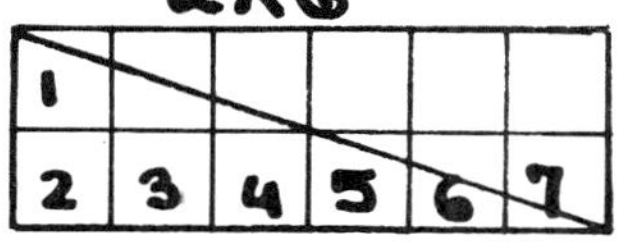

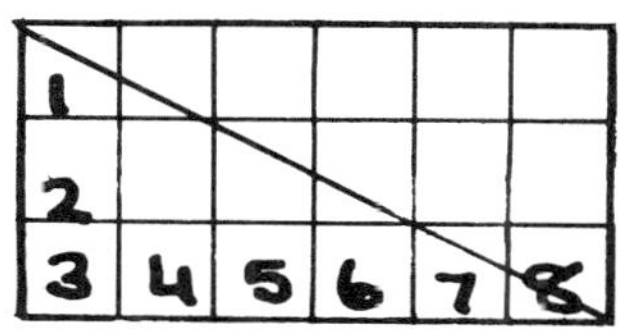

3X6

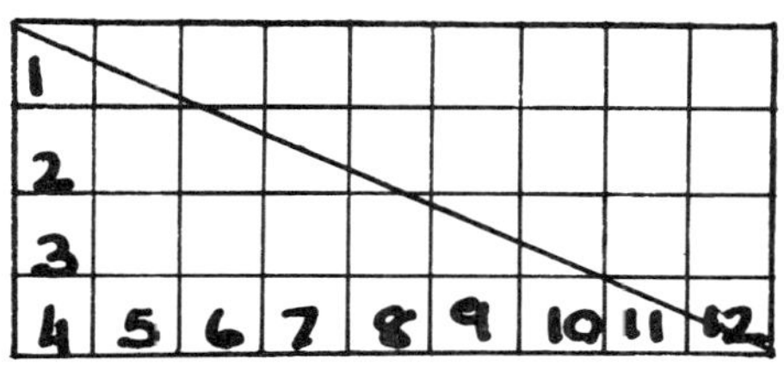

4X9

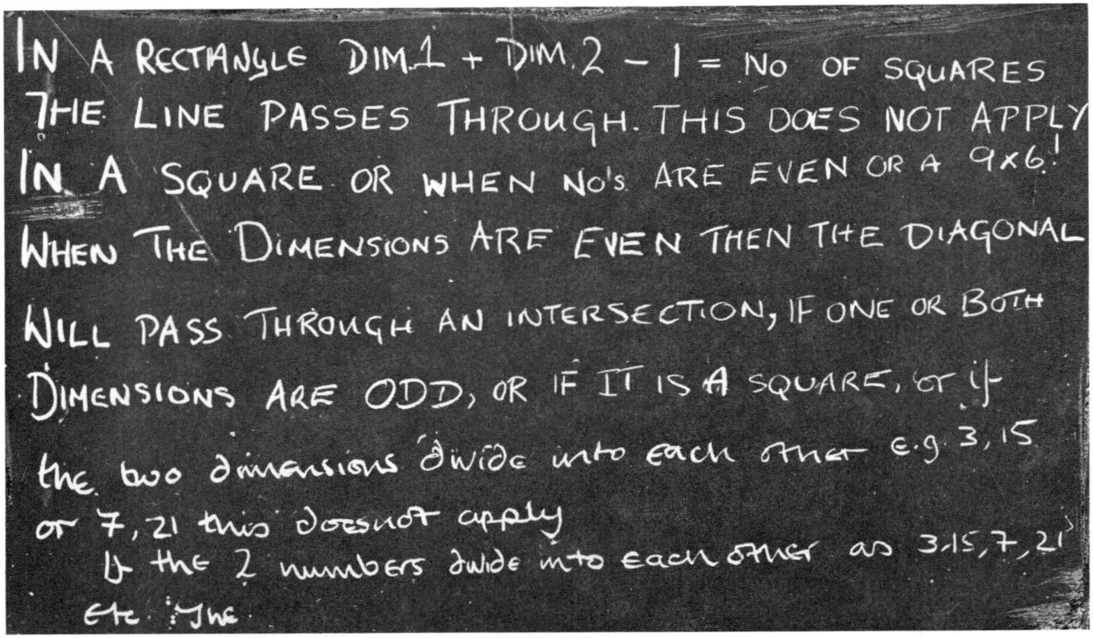
IN A RECTANGLE DIM.1 + DIM.2 − 1 = No OF SQUARES
THE LINE PASSES THROUGH. THIS DOES NOT APPLY
IN A SQUARE OR WHEN No's ARE EVEN OR A 9x6!
WHEN THE DIMENSIONS ARE EVEN THEN THE DIAGONAL
WILL PASS THROUGH AN INTERSECTION, IF ONE OR BOTH
DIMENSIONS ARE ODD, OR IF IT IS A SQUARE, or if
the two dimensions divide into each other e.g. 3, 15
or 7, 21 this doesnot apply
If the 2 numbers divide into each other as 3,15, 7, 21
etc. The

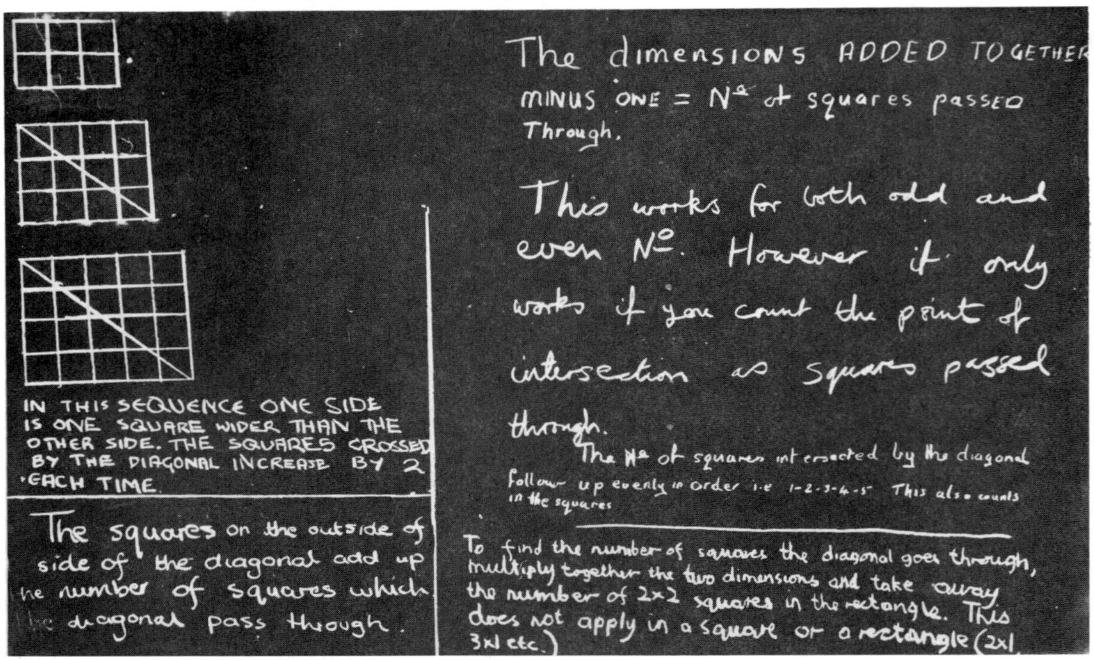
IN THIS SEQUENCE ONE SIDE IS ONE SQUARE WIDER THAN THE OTHER SIDE. THE SQUARES CROSSED BY THE DIAGONAL INCREASE BY 2 EACH TIME.
The squares on the outside of side of the diagonal add up he number of squares which diagonal pass through.
The dimensions ADDED TOGETHER MINUS ONE = No of squares passed Through.
This works for both odd and even No. However it only works if you count the point of intersection as squares passed through.
The No of squares intersected by the diagonal follow up evenly in order i.e 1-2-3-4-5 This also counts in the squares
To find the number of squares the diagonal goes through, multiply together the two dimensions and take away the number of 2x2 squares in the rectangle. This does not apply in a square or a rectangle (2x1 3x1 etc.)

It can be fruitful to compare one's own approach to the problem with that of the children. Investigation often begins with empirical observation. What determines the choices of the first $m \times n$ rectangles to be studied? It is perhaps unfortunate if you start with $m=n$ though then you not only quickly guess what the required number N is but perhaps also see why this must be so. Otherwise you might soon induce from your first few observations a formula $N=m+n-1$ that fits the pattern you are getting. How confident are you? Do you look for trouble deliberately? What happens if you now start drawing squares? Or perhaps you stumbled on a rectangle whose diagonal goes through a corner of one of the unit squares. Did you draw it accurately? What if you are satisfied that you did? What a nuisance! The formula does not seem to fit. But one does not give up a good scent so easily. How can it be made to fit?

You can formulate a theorem. But when is it true? You carefully describe the situations in which it is true. You make it true. Sometimes you even describe the situations as ones for which the theorem is true. Your theorem becomes a definition. Some scent is over-powering.

If and when you satisfy yourself that you have described when $N=m+n-1$, how confident are you now? Do you see why it must be so? What was the pupil who counted round the edges of the rectangle getting at? What about the awkward diagonals that go through corners? We may or may not have accommodated them into our framework but when does the awkwardness occur? Is there a single general formula that covers all rectangles?

There was work that considered half-squares (see p. 14). How can the problem be extended to $m \times n$ rectangles where m and n are not necessarily integral? What happens if one counts—as one boy did—all squares inside the rectangle that the diagonal passes through? Can the problem be extended in other ways?

the role of the teacher?

We have given a lengthy account of the children's work in order to indicate the sort of activity that we consider valuable and important. If the reader is impressed—as we are—by what the children did he will be interested in the ways in which it can be achieved. Certainly it is not just a question of introducing a class to some simply-formulated problem and leaving them to solve it. Such a procedure could produce very different results in different situations. Though the role of the teacher in this piece of work may appear minimal, it is clear that it must have been very important in the creation of the conditions in which it could take place.

What is the role of the teacher?

How can he create the conditions in which creative and independent work can take place?

A man once complained to the Hodja that there was no sunlight in his house.

'Is there any sunlight in your garden?' said the Hodja.

'Yes,' replied the other.

'Then put your house in your garden,' said the Hodja.

The heart of the matter lies in the nature of the relation between the teacher and pupil. Ultimately in the relation of the teacher to himself. Teaching soon becomes a moral voyage. But one in which the traveller has to choose his own map. We raise the issue here not to pursue it any further—there are plenty of available maps—but in order to suggest that, though we shall be dealing with other considerations, this is the centrally important one. 'The most important lesson-preparation is to prepare oneself'.

authority

One Friday, Nasreddin Hodja mounted the mimbar or pulpit in the mosques of Ak-Shehir to give the sermon.

'O true believers, do you know what I am going to talk to you about today?' he asked.

The congregation looked at each other in some surprise, then shook their heads.

'We have no idea,' they said.

'If you have no idea,' said the Hodja, 'what is the use of my talking to you?'

With that he descended from the mimbar and went home.

The following Friday he entered the mosque, mounted the mimbar and again asked the congregation:

'O true believers, do you know what I am going to talk to you about today?'

'Yes,' said the wily ones.

'Well, if you already know,' said the Hodja, 'what is the use of my telling you?'

And again he descended from the mimbar and went home.

When he again entered the mosque, he mounted the mimbar, and asked the same question:

'O true believers, do you know what I am going to talk to you about today?'

The congregation had long since prepared their reply.

'Some of us do, and some of us do not,' they said.

'In that case,' said the Hodja, 'let those who do tell those who do not.'

And away he went.

An arrangement of rows and columns like this can be particularly useful for examinations or individual work.

The same desks can be rearranged for group work to gain a sense of extra space and to enccurage communication.

Tables are more flexible units than desks. Various shapes and sizes are available and can often be ordered to your own requirements. The tables shown here are 120 × 60 × 75 cms.

Semi-Sweet Assorted

A variety of books should be available in a mathematics room. They are more attractive and more accessible when displayed on sloping shelves.

In the corner of this classroom, children's work can be stored or displayed on pegboard, on shelves, or in folder racks (made with hardboard and curtain wire).

When the teacher's desk is moved to one side the blackboard can be used by children to record their own work or to participate in the communal recording of a class discussion.

Whose classroom is this?

If every mathematics teacher has his own classroom then some care and attention can be given to the provision of a proper environment in which he and his pupils can work.

This is the way infants work . . .

What is the appropriate way for adolescents?

the classroom

The preceding photographs indicate various ways in which a standard classroom may be currently arranged. We have no blueprint for an ideal mathematics room. But thinking about the room one works in can be a useful starting point. Whatever the arrangement there can be no excuse for drabness. It is strange and sad that so many children move from a gay and lively junior classroom where their work is regularly displayed to secondary schools with bare walls and little pride in work or place.

The physical setting may reflect the way in which the teacher works. This can vary from the traditional situation where the whole class is working at the same topic to one in which each individual has his own particular task. It is not very helpful to see these as mutually exclusive alternatives. It seems to us, however, that at various times it would be wise for a teacher to take his own temperament into account, and to have some understanding of what he can cope with. For some it may be difficult to bear the anxieties involved in allowing children to work at different tasks. For others, it may be difficult to summon the authority and *self*-discipline required to enable a group to interact. At best the way of working will be dictated by the needs of the students and the environment appropriately arranged so as to be facilitating for the task in hand.

What is a suitable starting point for the teacher who decides on some occasion to encourage his pupils to work individually at separate tasks? A class that is familiar with such a way of working will find its own problems to investigate perhaps arising from previous work, from casual hints and suggestions, or from material left lying around. But it is much more difficult to start the process from scratch. Any romantic belief that it is simply a matter of leaving them to get on with it can be very quickly shattered even though leaving them to get on with it may remain a useful aim.

When the teacher arrives late for a lesson are people already working? Why not?

workcards

Workcards can be a useful way of getting things going though using them is not as simple and straightforward as it is sometimes made out to be. Many cards turn out to be assignment sheets through which the student slavishly works re-enacting situations in the compiler's mind and having nowhere to go when the last instruction is completed. Others may be too open for a particular student. Whatever the contents of the card there are various organizational details that can only be mastered by experience.

Anyone who is not used to having his pupils work at separate tasks on occasions may well be daunted at the prospect of setting up some work through workcards. But most teachers can easily spare one period a week or one whole week a term to experiment with new ways of working with a class. Once the first step has been taken further developments appropriate to the particular situation are easily introduced. We would recommend beginning with modest aims and an initial emphasis for the teacher on getting the feel of the organizational problems involved. It is because we feel that these are the most important difficulties that we give a brief account of a possible unit of workcards for a first step in this direction. We hope that the use of such cards would only be a temporary step.

Various points arise when it comes to preparing a set of workcards. They may provide a progressive sequence of work in which case they will have to be completed in consecutive order. Alternatively, they may not be in any sequence even though all about the same topic. This second method is probably the easiest to organize. In this case the cards are not conceived as a teaching unit. The student does not have to attempt every card in order not to miss a vital piece of work. In fact, a full investigation of one or two cards is preferable to a superficial look at several.

Some cards may be more challenging than others and more, or less, direction may be given as appropriate. The card may merely indicate a simple starting point which can be followed up with further suggestions which the teacher may have in mind, or which—as often happens—are raised by a pupil's initial work. In general, it should always be possible for all pupils to attempt some work from each card though clearly the levels of investigation will vary. To begin with it may be that the teacher decides when a card is finished. A class working in pairs will need 20 cards. These need not be all different; two copies of a set of 10 different cards is satisfactory. It will be worth taking some care to present the cards so that they are attractive and practical. The cards may be hand-written or typed. Postcards are often convenient but larger sizes may be preferred and these can be cut from thin card. Six-sheet card is suitable for typing. To prevent the cards from getting grubby or torn they may be placed in plastic wallets or covered with transparent adhesive plastic. They will, however, still need to be renewed fairly often, but this provides an opportunity to improve them, or to make additions and deletions to the set.

How does one get started?

One way is for the teacher to deal the cards out to the class bearing in mind individual needs and abilities. It seems preferable, however, to provide some choice at least for the first cards to be tackled. To do this all the cards have to be seen. They could be passed round or larger versions—to be copied down—may be pinned up around the room. It may be necessary to control the options for any card which requires special materials in cases where the supply is limited. Thus if a card requires a calculating machine it may have to be worked in turn. Pupils may work individually, or in pairs, but in any case it is always useful for them to discuss results with each other.

In the following five pages we give the contents of a unit of ten workcards on the theme of prime numbers. These could be used with mixed-ability eleven or twelve year olds.

DIAGONALS

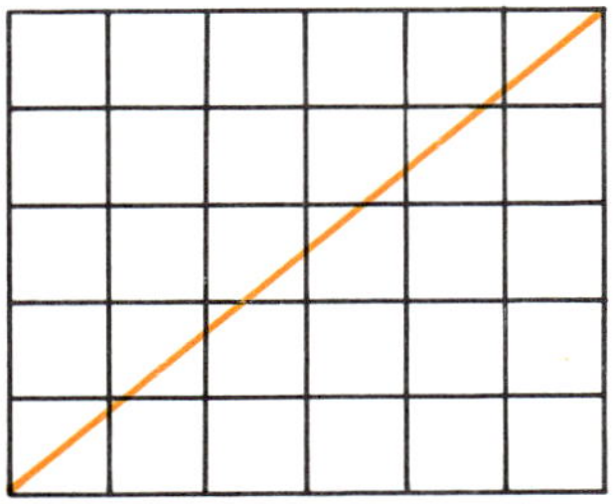

On squared paper draw a rectangle *five squares* by *six squares*.
Draw a diagonal of this rectangle.

How many squares does the diagonal pass through?

Do this for other rectangles.

Can you forecast the number of squares passed through if you know the length and width of the rectangle?

REBOUNDS

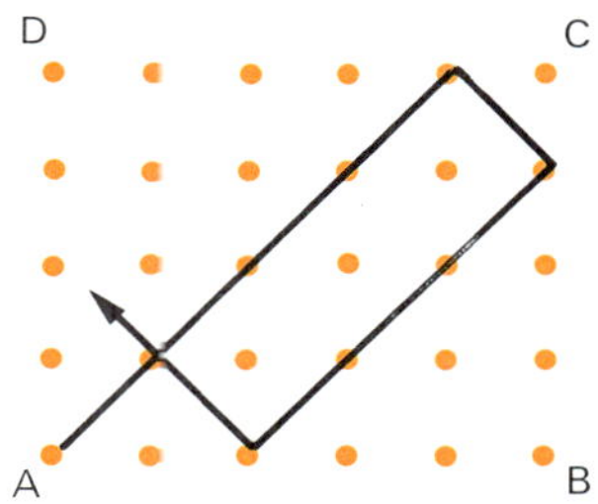

Make a rectangle of dots, *five dots* by *six dots*. Label the corners A, B, C, D.

Starting at A draw a line which goes diagonally across the squares until it reaches an edge. Turn through a right-angle and continue the line to the next edge. Go on doing this until you reach a corner.
Which corner did you reach?
How long is the line?

Investigate for various rectangles.

PRIMES AND SQUARES

$5 = 2^2 + 1^2$

$29 = 5^2 + 2^2$

Fermat, a French mathematician, proved that certain *prime* numbers are the sum of two square numbers. For example, the two prime numbers 5 and 29 can be written as shown above.

For each of the prime numbers less than 100, find out whether or not it can be written as the sum of two square numbers.

Can you find out anything about the prime numbers which are the sum of two squares?

CIRCLES AND POLYGONS

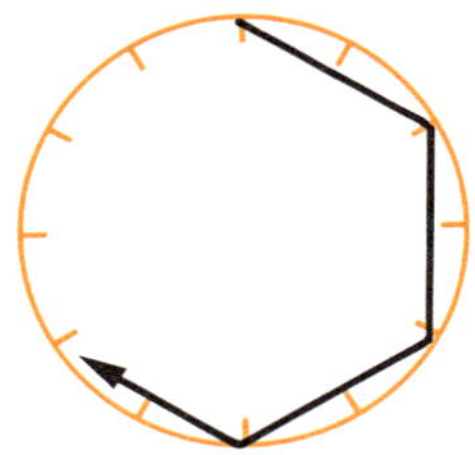

You will need circles with twelve points equally spaced on the circumference.
Starting at any point, join the points by straight lines *leaving two spaces* each time. Carry on until you reach your starting point.

What shape have you drawn?
What happens if 3, 4, or other numbers of spaces are left?

For which *numbers of spaces* do you reach the starting point after going only once around the circle? How many times do you go around the circle for other *numbers of spaces*?

SPIROGRAPH PATTERNS

You will need a spirograph.
Look at the *ring* marked $\frac{144}{96}$ and decide what these numbers represent.
Do the same for the *wheel* marked 36.
Using this wheel and ring make a pattern.

The pattern can be described as '8-pointed'.
Record the information:

teeth on ring	teeth on wheel	number of points
96	36	8

Make patterns using other wheels on the same ring, and also using the other ring. Record your results on the table.

What do you notice about the numbers in the table? Can you forecast the number of points if you know the number of teeth on the ring and wheel.

PRIME PATTERNS

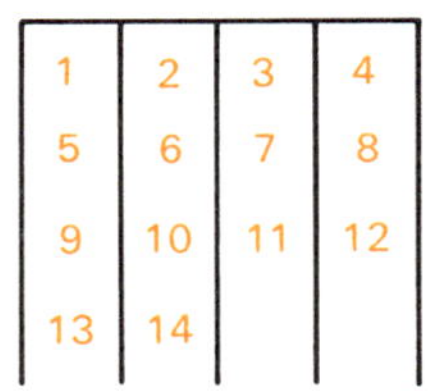

Write numbers from 1 to 100 in four columns.
Draw a ring around each *prime* number.
What do you notice about the primes. Can you explain why this is so?
Do this again using other numbers of columns.

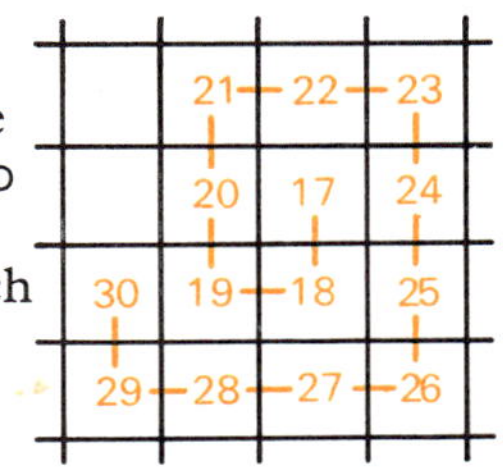

On squared paper write numbers from 17 to 100 in a spiral as shown.
Draw a ring around each prime number.
What do you notice about the primes?
Do this again starting the spiral with other numbers.

Experiment with other arrangements of numbers and see if the primes form a pattern.

PEGBOARD PATTERNS

You will need pegboard and pegs.
Place a white peg in the board.
Choose any hole and place a black peg in it.

(a) In some cases there will be a peg or an empty hole directly in line with the white peg. If so, change the black for a coloured peg.

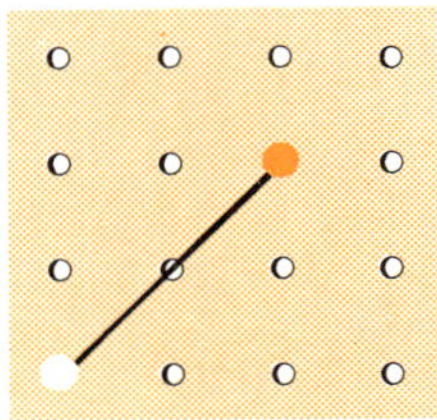

(b) otherwise there will be no holes or other pegs directly in line with the white peg. In this case leave the black peg.

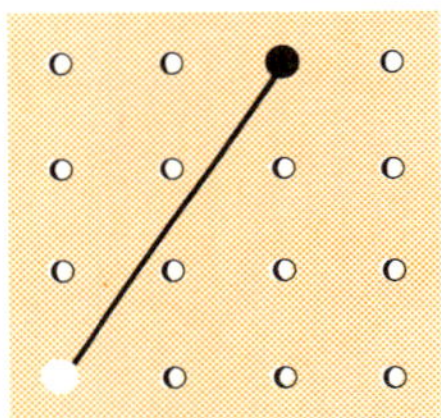

Place a peg in each hole of the board. Explore the patterns made by the pegs.

SUMS OF PRIMES

7 +		
19	2 +	
23	53	37 +
29	23	41
78	78	78

Choose four prime numbers less than 50. Find their sum.

Can you now get the same sum with three prime numbers?

Can you get it with two?

Write the numbers from 1 to 50 as the sum of prime numbers, using the *least number* of primes in each case.

DIVISORS

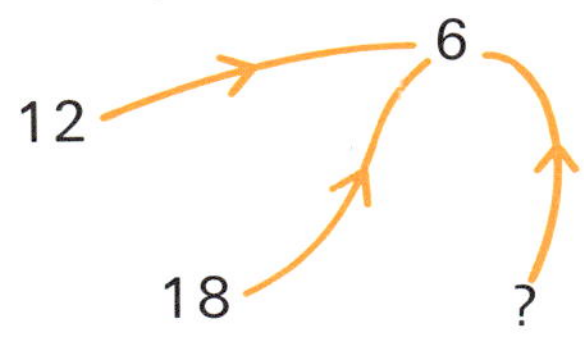

The set of divisors of 12 is:

{1, 2, 3, 4, 6, 12}

So 12 has six divisors.
This is shown on the graph above by an arrow from 12 to 6.
Check that an arrow should also be drawn from 18 to 6.
Can you find other numbers with six divisors?

Draw arrow graphs showing the number of divisors of other numbers.
Investigate the relation between numbers having the same number of divisors.

SUM OF DIVISORS

The set of divisors of 12 (*except 12 itself*) is:

{1, 2, 3, 4, 6}

The sum of these divisors $(1+2+3+4+6)$, is 16.
This is shown on the graph above by an arrow from 12 to 16.

By adding the divisors of 16 (*except 16 itself*) check that an arrow should be drawn from 16 to 15.

To which number should an arrow be drawn from 15? Continue the chain.

Investigate chains starting with other numbers.

write-up

Whatever the way in which work is initiated, some sort of record may be required. Now when a technique is being rehearsed, i.e. the student is working through exercises, then the solutions are written down as the problems are solved. But when a real problem is being investigated for the first time, there are many vague thoughts, tentative jottings or experimental actions, before anything crystallizes. And a solution may appear in more appropriate form than marks on paper; for example, a film, a particular wiring of a circuit, a pattern of tiles, and so on.

When should a 'write-up' be demanded?

What form should this write-up take?

An unfortunate convention usually decrees that any scientific work should be described impersonally and should make no reference to the way in which results were achieved. This convention has been increasingly questioned in recent times but in any case we see no reason why it should be prematurely forced into schools. As indicated in our account of some work done by pupils, we feel it is valuable and natural for them to describe process as well as product. It takes much patient care to encourage and enable them to do this. Older students may already have adopted the adult conventions. On the other hand, younger children may need weaning from the descriptive personal (*I did this . . . then I did that . . .*), though this is perhaps better than the experimental passive (*This was done . . . then that was done . . .*).

A write-up—and incidentally we think this is a good word to use with pupils—may be initiated after some activity merely by reminding the class what material is available, and asking them to show in some way what sort of things they have been doing. In some cases it may be preferable to suggest some ways of going about this (*Try to explain to someone else . . . make up some problems . . . how did you start? . . .*). In any case, it is obviously important to discuss the nature of such work from time to time, and to accept encouragingly, but not uncritically, early efforts.

How and where is work kept?

Who reads it? How is it assessed?

assessment

It is all too easy to become dominated, indeed obsessed, by problems of assessment. Marking and correcting as such are neither right nor wrong but part of a dialogue. Children often want to know what the teacher thinks of their work and this can often be conveyed more sensitively by written or spoken comments, whether grades are awarded or not. We observe that when there is a real challenge there tends to be a creative and personal response. And as a teacher once remarked, 'One becomes less able to give D's'.

Two men who were involved in an argument appealed to the Hodja to settle it for them. The first man gave his version of the quarrel, and when he had listened attentively to all he had to say, the Hodja nodded his head, and said:

'You are right'.

'It is not like that at all!' exclaimed the second man, and began to relate his side of the story to the Hodja. When he had finished, the Hodja said:

'You are right.'

The Hodja's wife, who had listened to everything, could bear it no longer.

'Hodja,' she said, 'they cannot both be right!'

The Hodja thought for a minute.

'Wife, you are right, too! he said.

an individual write-up

Here is an example of an individual write-up—an account by an able eleven year old boy of an investigation of networks.

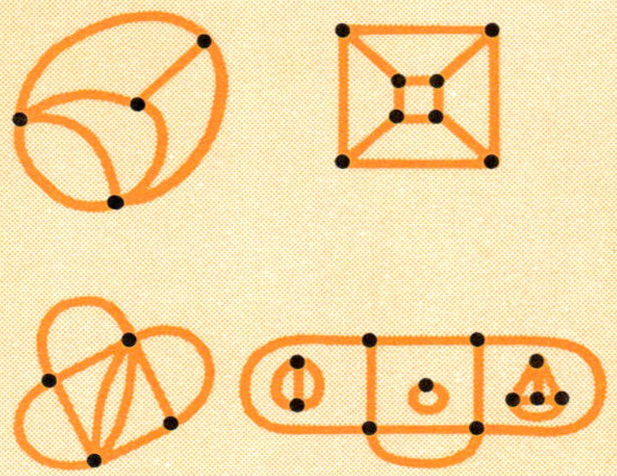

A *network* is a set of lines, junctions and spaces, which compose a shape.

There are two kinds of network, connected and disconnected. First I will consider the connected networks. There are several rules for these such as:

—where two or three lines meet you have a junction.

—if a straight line has a junction in it, it is counted as two lines.

—a circle has to have a junction on it.

We have to have a method of recording this type of network, so I decided to record the number of junctions, the number of lines and the number of spaces. When counting the spaces you had to record the 'outside' space.

Some networks are shown with their 'numbers' underneath.

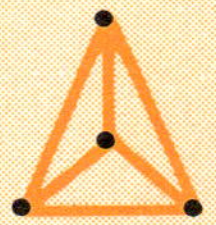

L J S
6 4 4

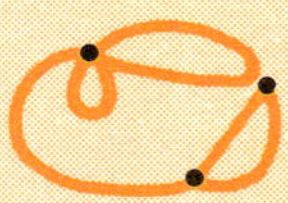

L J S
5 3 5

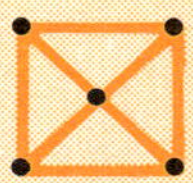

L J S
8 5 5

I then attempted to discover if, from the lines and spaces, I could discover the number of junctions.

I took this network . . . and removed one line.

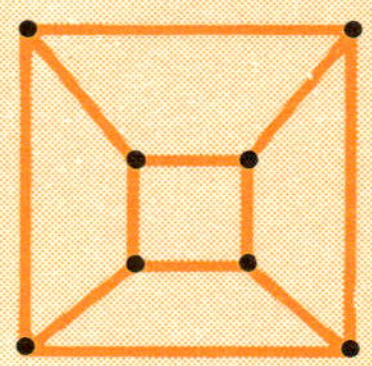

L J S
12 8 6

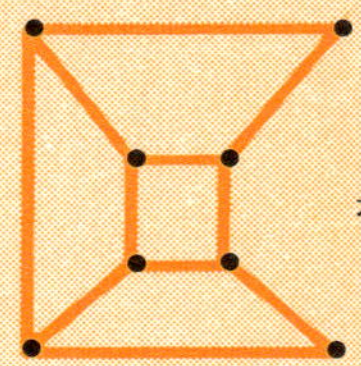

L J S
11 8 5

I continued doing so without destroying any junctions, that is until I got this shape . . .

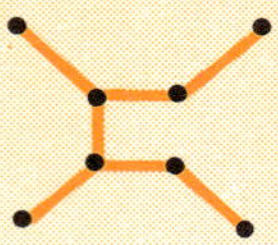

L J S
7 8 1

L	J	S
12	8	6
11	8	5
10	8	4
9	8	3
8	8	2
7	8	1

In doing this I got a table of results.

Now, if you add L and S together you get 18 (in the first 'number'). From this you wish to get J (8). To get it you need to take 10 (in the first case) away. To arrive at 10 you take 6 (S), subtract 1 and multiply by 2. In a formula: $J=(L+S)-((S-1)\times 2)$.

This was simplified like this:

$(L+S)-((S-1)\times 2)=(L+S)-(2S-2)$
$=(L+S)-2S+2=(L-S)+2$

The formula works for all cases $(L-S)+2$ and adaptions work to find S and L: $L=(J+S)-2$ and $S=(L-J)+2$.

This was first discovered by Euler, but it works only for *connected* networks.

The formula $(L-S)+2=J$ does not work for disconnected networks, so I decided to find a formula which did. When recording a disconnected network you have to record the number of parts (N).

Some disconnected networks and their 'numbers' are shown.

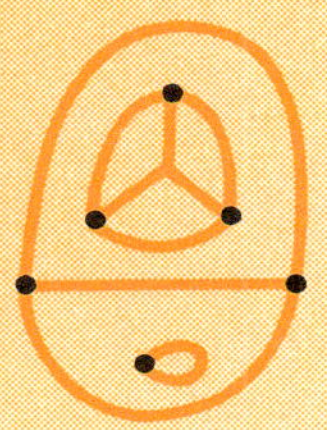

L	J	S	N
10	7	7	3

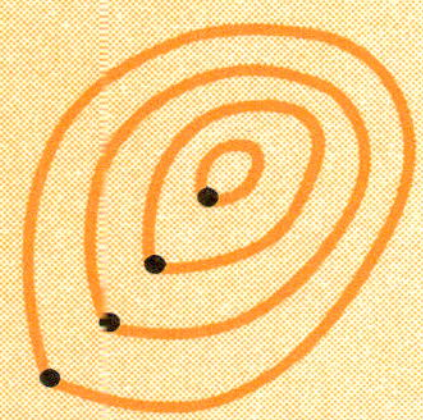

L	J	S	N
4	4	5	4

L	J	S	N
9	6	7	3

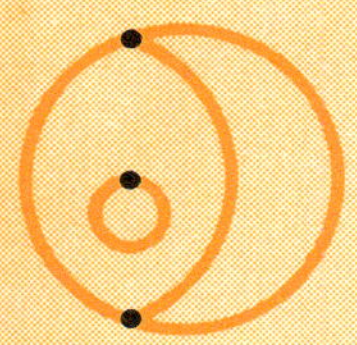

L	J	S	N
4	3	4	2

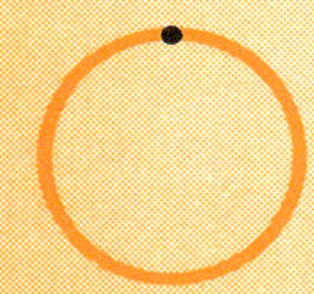

L	J	S
1	1	2

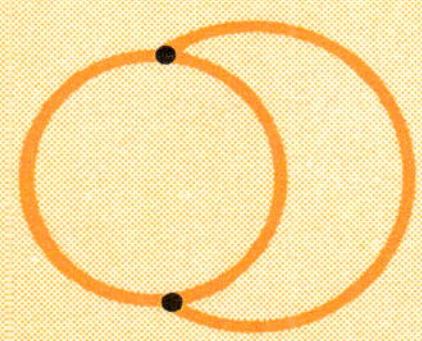

L	J	S
3	2	3

I first turned the following network into its 'components'.

If you add the numbers 1 1 2 and 3 2 3 together they come to 4 3 5. In the first number (leaving 'N' out), L J S is 4 3 4. This is because when you take the two figures separately you count the outside shape twice but when they are together you only count them once.

If we go back to the formula $(L-S)+2=J$, it shows that whenever you use 'S' you must add $N-1$ onto it. (Every part except the outside one has not got its 'outside' counted).

This gives us the formula: $(L-(S+(N-1))+(2\times N)=J$.

Which is simplified like this:

$$(L-(S+(N-1))+(2\times N)$$
$$=(L-(S+N-1))+(2\times N)$$
$$=(L-S-N+1)+(2\times N)$$
$$=(L-S-N+1+2N)$$
$$=L-S+N+1$$

Adaptions work for L and S:

$$L=J+S-N-1 \qquad S=L-J+N+1$$

This formula is different to Euler's because it works for all networks, connected or disconnected.

a joint write-up

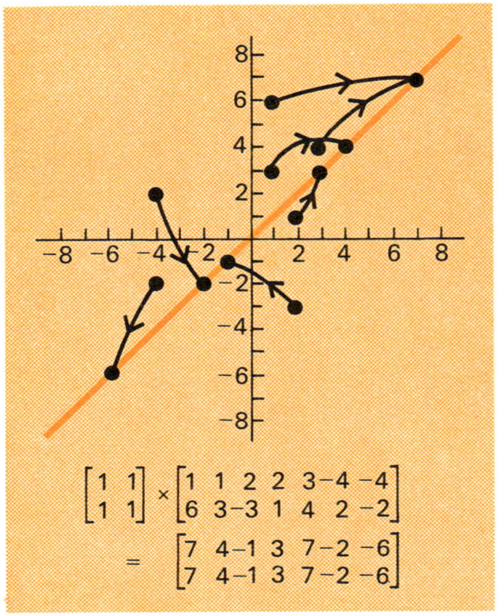

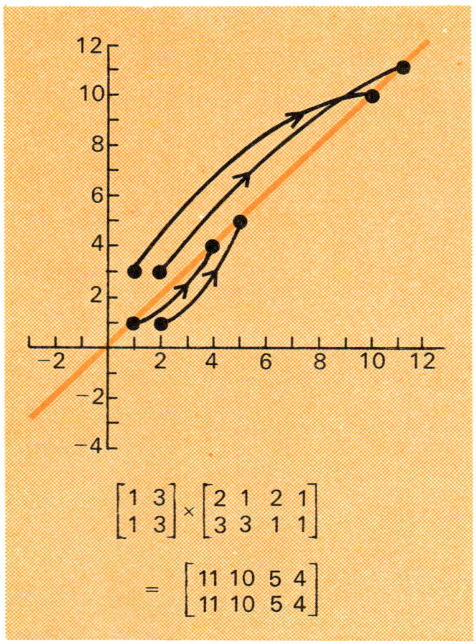

Here is a write-up by two fourteen year olds of average ability :

We took points on the plane (which were dotted all over the place) and then we took a matrix. We multiplied the points by the matrix and transferred the answers back on to the plane to see where the transferred points landed.

When we had transferred the points we found that they all ended up in a diagonal line. We thought that perhaps it was because the matrix had all the same numbers in it. So we tried some other matrices with all the same numbers in them. These all made a diagonal line as well. When we had done this we thought we had finished the work on the diagonal line but we found a different matrix that made the same diagonal line. This matrix was $\begin{pmatrix} 1 & 2 \\ 1 & 2 \end{pmatrix}$

When we looked at this matrix and one of the other matrices we found they had one thing in common, that was that the top and bottom rows were both the same. So we tried another matrix to see if it worked. As you can see by the diagram this worked as well. After this we said that all matrices, no matter how high their numbers are, so long as both the rows are the same, make any point into a diagonal line. (This line goes from top right to bottom left). The matrix can have negative numbers in it, and so can the points be negative.

After we had made this statement we tried to get a matrix that made a diagonal line that went from top left to bottom right. We thought that the matrix would contain a mixture of positive and negative numbers, as the others had either one or the other but not both. The first matrix we tried was $\begin{pmatrix} 1 & 1 \\ -1 & -1 \end{pmatrix}$ and we found that it worked. We tried doing it with the negative numbers on the top. This worked as well. So we now have a way of telling what the matrix could be if we wanted the points to end up in a diagonal line where (if we took away the negative sign) the □ and △ numbers are the same.

Now that we had found the matrix for diagonal lines we wanted to find the matrix for straight lines (horizontal and vertical). While we were trying to do this we found another matrix that made a diagonal line only the line was at a

different angle than the one before, e.g. the one before was $45°$; this one $22\frac{1}{2}°$.
We also found that when we reversed this matrix it gave us a diagonal line at $67\frac{1}{2}°$.

We now had four different diagonal lines. What we then did was to get the last two diagonal lines to turn and go from top left instead of top right. We found the two matrices which made the lines and drew the lines on the diagram.

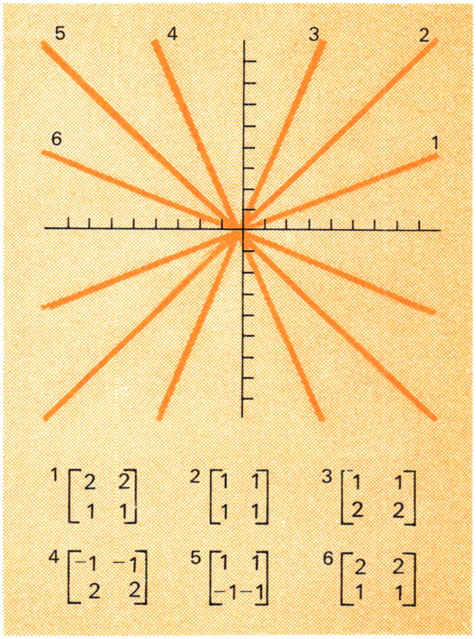

This diagram shows all the diagonal lines we have found so far.

When we looked at the matrices that were used for the lines (excluding the first one) we saw that from one dot to the other the line went across 2 and up 1; for the matrices $\begin{pmatrix} 2 & 2 \\ 1 & 1 \end{pmatrix}$ and $\begin{pmatrix} -2 & -2 \\ 1 & 1 \end{pmatrix}$ and another line went across 1 and up 2 for the matrices $\begin{pmatrix} 1 & 1 \\ 2 & 2 \end{pmatrix}$ and $\begin{pmatrix} -1 & -1 \\ 2 & 2 \end{pmatrix}$

When we saw this we thought that if we made a matrix like $\begin{pmatrix} 3 & 3 \\ 1 & 1 \end{pmatrix}$ the line would be across 3 and up 1 and if it was $\begin{pmatrix} 3 & 3 \\ 2 & 2 \end{pmatrix}$ it would go across 3 and up 2. We tried this to see what happened. We found we were right. We now said we knew all there was to know about diagonal lines. We then went on to horizontal and vertical lines. To get a horizontal line we used any matrix with $(0 \; 0)$ in the top row and to get a vertical line we used any matrix with $(0 \; 0)$ in the bottom row. The only thing wrong with this was that the horizontal and vertical lines we made came on the number line. This made it difficult to see the line we made against the number line. We then tried to see if we could get a horizontal and/or a vertical line which was away from the number line. We have not yet found any matrices that do this and we don't think there is one.

a first-form write-up

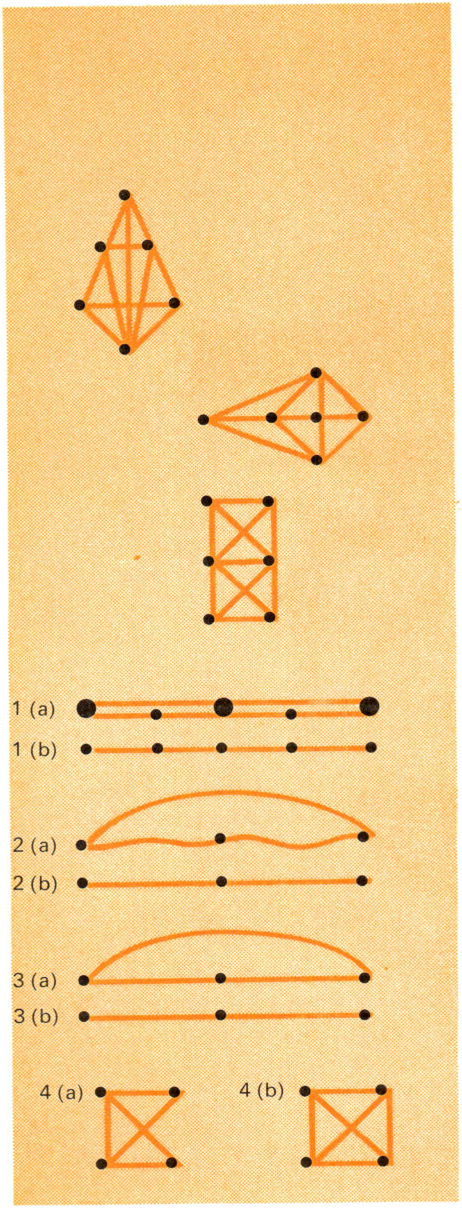

After a week's work arising from the starting point described on p. 139, a first form was invited to produce a joint record of what they had done, individually and in groups. The class discussed the way this could be done and eventually a list of contents emerged with different aspects assigned to particular people. After a false start it was realized that there would also have to be agreement about page format for the whole to be presented as a single pamphlet. We are grateful to the authors for permission to quote from this pamphlet:

Dots and direct routes

INTRODUCTION

Lots of people went up to the board and drew 6 dots in different patterns.

When we had got quite a few, we joined each separate diagram up from dot to dot with straight lines.

We found that although there were the same number of dots there was different number of lines.

We made up a few more different diagrams and found that however hard we tried a certain number of lines were impossible (or we couldn't find them).

We couldn't understand this, so we tried another number of dots, so this is how this project came about.

RULES

1. All dots must be the same size.
Not like this! (1a) Like this! (1b)

2. All lines must be straight.
Not like this! (2a) Like this! (2b)

3. If there are three dots in a row, the two end dots cannot be joined together by one line. This means that if there are three dots in a row, there can only be two lines.
Not like this! (3a) Like this! (3b)

4. Each dot must have as many lines coming off as one less than the amount of dots. E.g. 4 dots, 3 lines coming off each dot.
Not like this! (4a) Like this! (4b)

5. There are three dots in a row, and another dot is placed just above the other end dot as in the diagram. We can call the dots A, B, C and D as in diagram. If a line is drawn from B to D then there *must* also be one from A to D. This rule is mentioned because it is possible to give the impression that B is blocking a line from A to D as in (second) diagram.

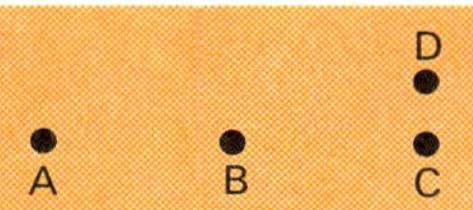

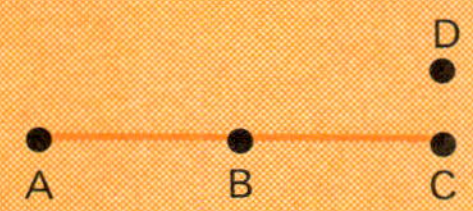

THE LAW OF MAXIMUM LINES

This rule tells you how to find the maximum number of lines, from a certain number of dots. The law is NUMBER OF DOTS × NUMBER OF DOTS − NUMBER OF DOTS ÷ 2, e.g. (6 dots × 6 dots) = 36 (− 6 dots) = 30 (÷ 2) = 15, so 15 is the maximum number of lines for 6 dots.

Here is a diagram showing 15 lines.

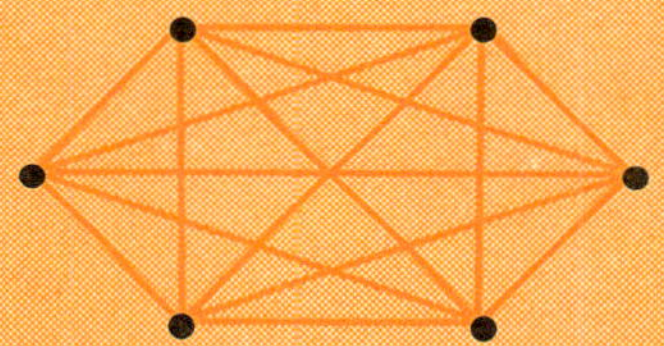

THE LAW OF MINIMUM LINES

This rule tells you how to find the minimum number of lines from a certain number of dots. The law is NUMBER OF DOTS − 1 = NUMBER OF LINES, e.g. 6 dots − 1 = 5 lines so 5 is the minimum number of lines for 6 dots.

Here is a diagram showing 6 dots with 5 lines.

LINES AND HOW YOU GAIN THEM!

If you have a line of dots, say 6, and you draw as many as you can between them you will get 5.

This is the minimum amount of lines you can get. If you put them in any other order you will get far more lines, e.g. . . . The next least number of lines is if you put 1 dot out of place. Then the next least is if you put 2, 3, 4, and 5 dots out of place.

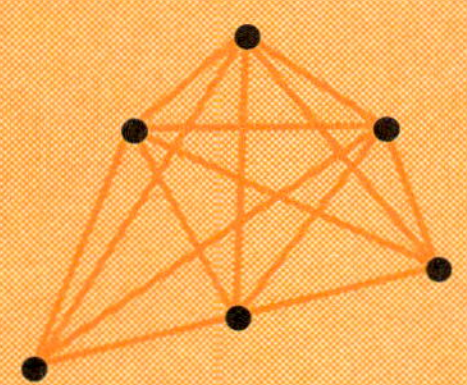

When you're putting the dots out of line, you must put them in line with another dot, or you will get more lines.

In this diagram the two dots out of the line are in line with one other dot, and there are 11 lines.

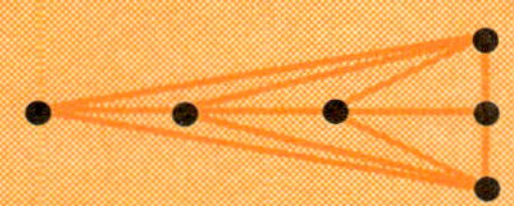

In this diagram the two dots out of line aren't in line with one other dot, and there are 12 lines. . . .

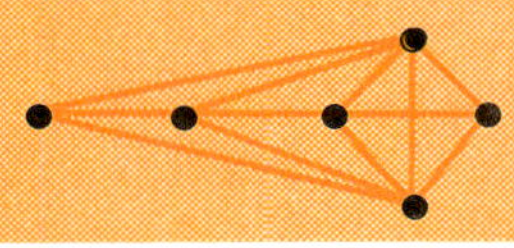

When you move one dot out of line you get 5 lines (using 6 dots) if you move two dots out of line you get 9. So all the numbers of lines between 5 and 9 you can't get. Because 5 is the least amount of lines and 9 is the next least amount. All the lines between 9 and 11 are impossible also.

write-up—continued

THE ELASTIC LIMIT

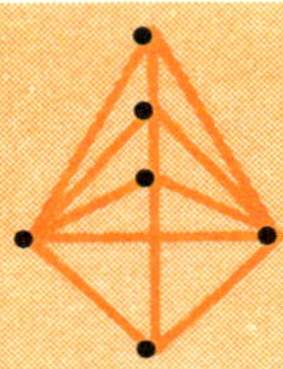

There is one thing that breaks this law of having the least amount of lines when you have them in line with another dot, and also of finding what lines can't be done. This is when you have a diagram like this . . .

In this diagram you get 12 lines.

LINES AND HOW YOU LOSE THEM!

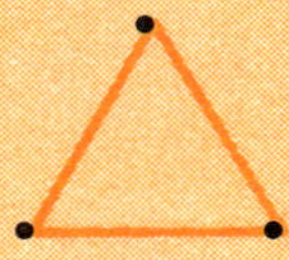

On the previous page you were shown how to gain lines. This page shows how to lose them.

If you take a three dot diagram and put it so that you get the most lines possible it looks like this.

The amount of lines is three. To reduce it to two lines this is the sort of thing that must be done.

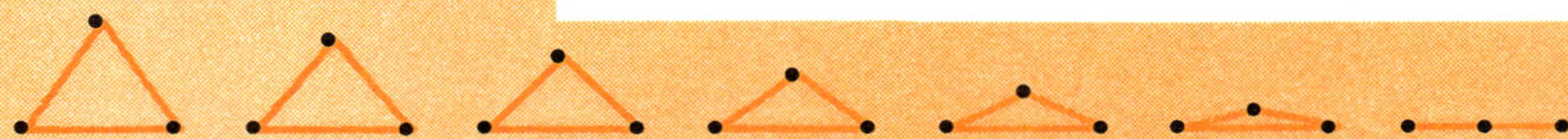

The seventh square holds the reduced lines and 3 dots. The first contains the most lines you can get. Here is the same sort of thing only with 6 dots.

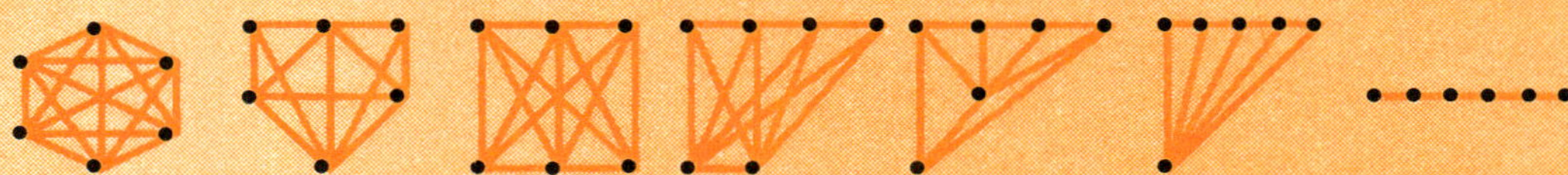

The law connected with this is that if three dots are put in a row the amount of lines decreases by one.

Here follows a number of diagrams showing the number of lines for various numbers of dots

I DOT No direct line possible.

(The work continued with complete diagrams and captions in the cases of 2, 3, 4, 5, 6, 7 and 8 dots which we do not reproduce here. This work was summarized in the next section):

	min	max	missing cases	
1				
2	1	1		
3	2	3		
4	3	6	4	
5	4	10	5 – 6	
6	5	15	6 – 8	10
7	6	21	7 –10	12–13
8	7	28	8 –12	14–16

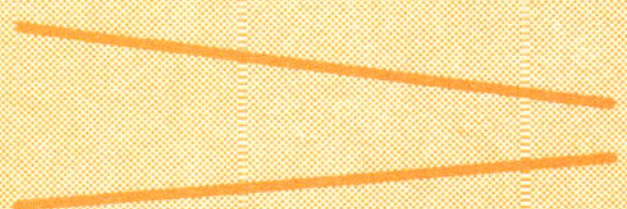

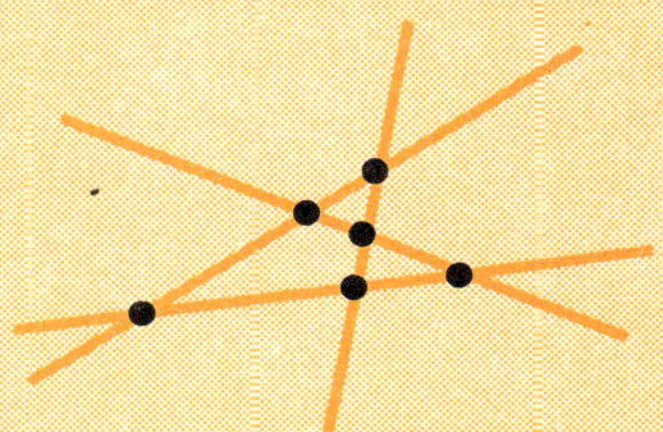

LAWS PROBLEMS AND UNANSWERED QUESTIONS

1. There is always a jump in the number from the minimum to the maximum (excluding 2 dots).

2. The minimum number of lines is always one line less than the number of dots (excluding 1 dot).

3. $\square \times 2 - 4 =$ number of diagrams (lines) not possible. ($\square =$ number of dots.)

4. No. dots × No. dots − No. dots ÷ 2 = Max. No. lines.

Unanswered questions:

1. Will a third group start carrying on from 8 dots? (See asterisk in table)

2. Could one be able to tell the max., min. (No. of lines) and how many groups?

REVERSING THE PROBLEM

Reversing the problem, how is it done?

To reverse the problem you first draw lines and where the lines cross you put a dot, this is called a point. As shown . . .

The example shows two lines and one point, the orange dot representing the point. This problem is reversed because on the previous pages we had a certain number of dots and these were joined together with lines, but in this problem we have a certain number of lines, none of which are allowed to be parallel and where they cross we put a point.

Rules:

1. Lines must not be parallel.

2. If two lines are slanting towards each other they must eventually cross.

3. One point only may be placed where the lines cross.

4. Lines must be straight.

(*Some diagrams and a table were then included. This section and the whole pamphlet ended with the following sentences*):

Owing to lack of time there are many unsolved problems. Perhaps you would like to have a go at them yourselves.

syllabus

Although teachers have much freedom within their own classrooms, most work within certain external frameworks. In the first instance, they will probably be part of a mathematics department which—either corporately or through the head of department—will determine certain ways of working. Inevitably, the most dominating constraint will seem to be the syllabus with associated text-books and external examinations. We think this framework could easily become more flexible in the future. But the unfortunate distinction between a 'modern' and 'traditional' syllabus will remain for some time. We do not intend to rehearse the well-worn arguments in favour of a reform in the mathematical content of a syllabus. But we observe that any attempt to restrict the range of mathematics that might be made available to pupils would be impossible within any active process of education in which anything must be grist for the mill. Any natural investigation of problems that arise will inevitably cover many prescribed topics of a syllabus. We are not in this context particularly concerned whether pupils should be prescribed some nineteenth century group theory or some seventeenth century calculations. The real choice is between the quick and the dead.

The shorter the written syllabus, the more alive its work can be. Apart from anything else, any topic must really start where the pupils are, and end where they leave it. It can be useful to save ourselves the self-deception of assuming that this can be closely determined beforehand. Similar remarks could be made about many text books though there are obvious ways in which they can be useful tools. Complete sets cost a lot of money that could be spent on other things. Moreover they can over-determine future developments. Where some alternative to working solely from a text book is wanted, adequate duplicating facilities become very important. It may also be worth considering buying only two copies of a book whose pages can then be torn out and mounted on card or stapled in small folders.

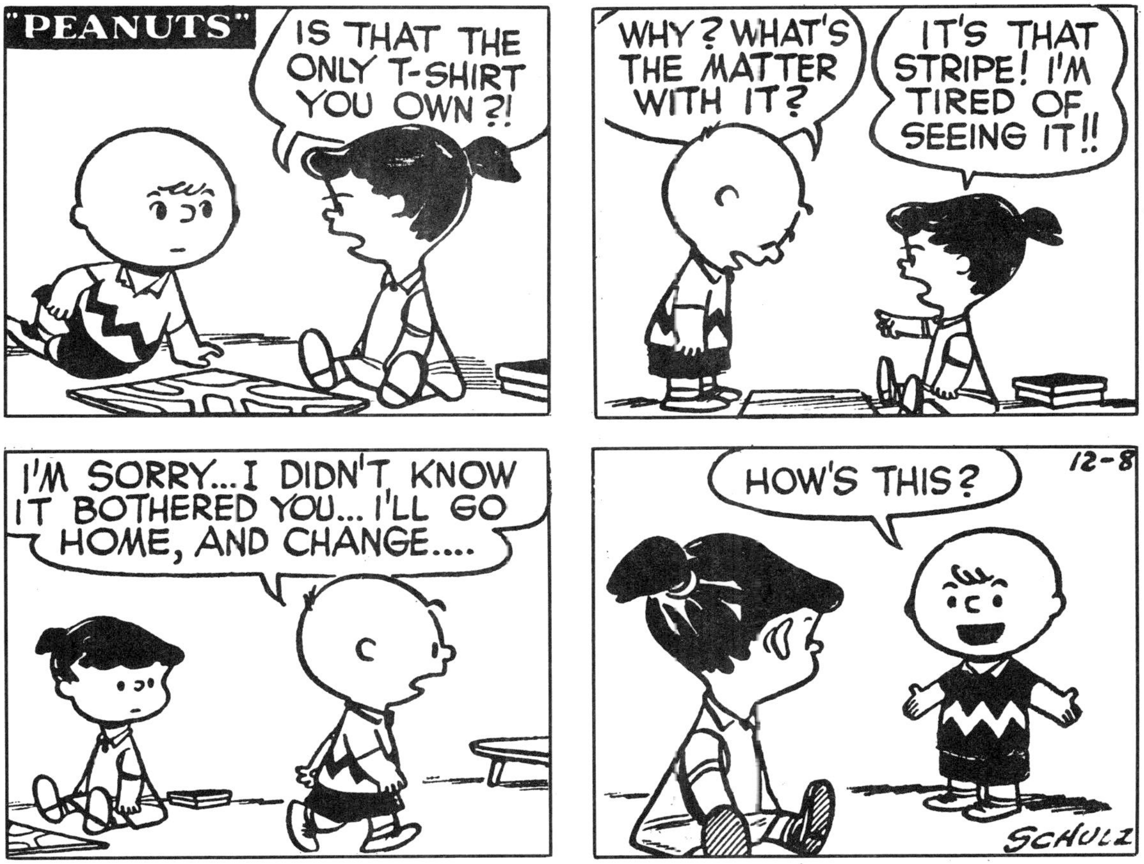

the new maths?

examination

As if all these possibilities do not already imply enough work for the harassed teacher, he is likely to become increasingly involved in external examinations. The Certificate of Secondary Education, in particular the mode 3 which permits schools to set and assess their own examinations subject to moderation, may well set a general pattern for the future. It is not clear why more schools have not taken up the option of the mode 3 examination. In some cases it is because it means too much work—and we do not find this unnecessarily unreasonable. But what is a teacher to do if he initiates certain flexible ways of working with lower forms when the time comes for them to prepare for and take public examinations? One solution is for the teacher to share his problem with his students, inviting them to down tools, as it were, and play another game, that of passing public examinations. In effect this may mean that students will have to change their style of writing-up work. For example, instead of the work quoted on p. 42, more evidence of efficient technique in manipulating matrices may be expected and required. This will involve a change of gear. But the time may come when this absurdity is seen for what it is. Students may prefer to go on working in certain ways, producing material they wish to examine and see examined.

Specially prepared projects and selected pieces of course work can be submitted for examination with some C.S.E. mode 3 schemes. These may also allow variable time limits to set papers. There are many difficulties involved in all this, and—as we suggested when discussing workcards—it may be important to start with modest aims, in order to cope with the organizational problems involved. For example, a hybrid scheme may be agreed with the examination board, whereby candidates take one of the board's papers and one set by the school. The marks for the latter may include a percentage allocated to a project.

Further information and examples may be found in the Schools Council Bulletins, and we also refer the reader to the A.T.M. report on examinations and assessment (see p. 64).

projects

A starting point which begins in class and is followed up immediately, allows a student to engage in a particular problem and pursue it in the next few lessons. The projects discussed here are intended for work over a longer period—a few weeks or even a whole term—and allow for a wider treatment of a topic. Such work is thought of as being completed outside the classroom, with regular occasions provided, say once a week, for individual consultations and discussions.

It may be desirable to give some guidance on the organization and presentation of the project. This will depend on the experience of the pupils (and the teacher). The project might be presented as a booklet, a collection of models with notes, some charts, a set of filmslides with tape, a filmloop.

In any case it will be important to have access to appropriate materials and a good reference library.

Teachers who wish to have course work taken into account in the C.S.E. can negotiate with the examining board to have a mode 1 paper complemented by a submitted project which can earn a certain proportion of the total marks. As part of an examination a project will have to be assessed in some way. Criteria for such assessment might include evidence of research and original work, mathematical content, presentation, and ability to discuss the contents of the project.

Topics for projects may arise naturally from work that has been going on in the classroom or may be initiated separately. In many cases they will involve a certain amount of bringing together of previously explored ideas as well as some further explorations. The extent to which a project may become merely a copying out of information from a reference book depends on the teacher. For some pupils, the activity of selecting and organizing the available information may be a very important and valuable activity but there are obvious dangers in choosing certain standard topics which are fully written about in various books. It is through regular discussion of the work in hand that the teacher can discern the appropriate challenge for a particular individual.

Clearly there should be a wide choice of possible subjects available. It may be useful to have duplicated lists of titles with brief indications of what might be considered in each case. Sometimes these notes merely serve to remind the student of some notions that will already be familiar. More specific suggestions could be prepared if these were required. Thus, to take the first item mentioned below, a special sheet of suggestions for an investigation of the Fibonacci sequence may include:

In the sequence 1, 1, 2, 3, 5, 8, . . . each term is the sum of the previous two. Continue the sequence.
Study the appearance of certain kinds of numbers, e.g. square, prime or odd numbers.
Look at the sums of any number of consecutive terms.
Explore the ratios of two successive terms.
Form the sequences in various finite arithmetics.
What patterns arise?
Make similar sequences starting with two other numbers.
. . .

We give a list of a dozen items that have been used as starting points for C.S.E. projects. There are plenty of other possibilities.

Fibonacci sequence Law of formation, patterns, general term, golden mean, link with Pascal's triangle, natural growth, applications to architecture, Wythoff's game, the sequence in finite arithmetics, generalizations, historical remarks, . . .

The cube Nets for model making, surface area, volume, diagonals, dissections, colouring faces, perspective drawings, routes on a cube, cubic numbers, boxes, . . .

The hexagon Regular and irregular, angles, lengths, symmetries, tilings, dissections, hexagon grids and lattices, crystals and honeycombs, games, hexagonal numbers, construction, applications, . . .

Envelopes and loci Curve-stitching, paper folding, films, spirograph, three dimensional, applications, detailed study of particular cases, historical, . . .

Polyominoes How many . . ., tilings, comparison with nets of solids, generalizations to other shapes, three dimensional forms, games and puzzles, colourings, . . .

Paper folding Geometry of paper folding, plaiting, linkages, folding strip problems, knots, bands, origami, parcels, paper aeroplanes, . . .

Shape in nature Flowers, crystals, fossils, tilings, growth series, spirals, weight–volume–surface area ratios, golden mean, astronomy, molecular structures, mudcracks, . . .

Packings Nets of solids, space filling, cuboids to hold given volume, volume and surface area requirements, container requirements, package design, examples, . . .

Dominoes Numbers in various sets, closed chains and unicursal routes, generalizations to polyominoes and to other shapes, tilings, games and puzzles, . . .

Calculating devices Abacus, slide rules, Napier rods, logarithm tables, nomograms, calculating machines, computers, programming, numerical work, history, . . .

Statistical investigations There are many situations providing material for a project on statistics. Most topics can be approached in the following way: collection of data (questionnaire if necessary), tabulation and representation, observation and calculation (averages, indices), conclusions and applications.

The following list might be helpful: public transport, traffic flow and control, newspapers, cars, local theatre, car parks, personal measurements, the family, the library, 'Which' survey, estimations, sport, your class, opinion poll, . . .

Other possibilities If there is some other topic you would like to choose make sure that you discuss it with your teacher before beginning work.

shared teaching and observation

When the C.S.E. examinations were first introduced, many teachers found that they were meeting and discussing with other teachers for the first time. If this seems surprising it is even more remarkable that many teachers working in the same department in a school have little idea of what goes on in each other's classrooms. The current uncertainties about syllabus—or indeed about the whole organization of schools—can usually focus attention on ways of working. The staff of a department may begin to meet regularly as a result of some particular discussion, say, on streaming.

One of the most significant changes in recent years has been the breakdown of the teacher's traditional isolation in his classroom. It is becoming more and more possible for teachers to work together in each other's rooms or indeed in each other's schools, though there are many anxieties and self-deceptions to overcome. Shared teaching and observation is rewarding for there can be a gain of insight into what is going on. And we must clutch at the smallest such gain because we know so little about what are very complex events. We emphasize the complexity of classroom interaction because it can so easily be denied in the interest of some simplifying, but mistaken, objectivity. Shared observations which are frankly understood to be subjective and participatory are a useful way of continuing discussion for the people involved.

Records of such observations can be helpful. These may vary from films or tapes, sometimes generally available like those of the Madison project in the U.S.A., to occasional written records published in journals and newsletters (notably those of the Association of Teachers of Mathematics), or circulated privately. The staff of a mathematics department may decide to maintain an accessible file of such observations as well as other material and information that can be shared. Thus it can be very useful to have a filing cabinet containing duplicated material and stencils, lesson notes, work sheets and so on, which all members of staff can use.

In the following five pages we give a selection of items from a logbook—a file of occasional pieces of writing about work done in a variety of schools. Such accounts may be hastily written notes at the end of the day or composed later after reflection and discussion.

extracts from a logbook

CUBES. There had been discussion about solids and faces (flats), corners and edges. Having decided a cube could be made from six squares, the class were given the task of finding how many different ways six squares could be arranged.

The children proceeded, in groups of between four and six, to find as many arrangements as possible. The first lesson had the groups arguing about 'which were the same arrangements'. Only on few occasions was I called in to settle an argument. During this first lesson there was little or no contact between groups.

The second lesson proceeded as before, each group sticking their arrangements of paper squares on a large sheet of paper. People began asking other groups how many they had found. I decided to keep well out of the way whilst this was going on. One boy shouted from the back that he had them all. 'How many', I asked. '35'. The boy's group checked on these and eventually decided that they had them all.

The peculiar thing about this was that the other groups decided to make 35 their target, accepting this as the lot. As the groups were getting about 30 arrangements there was great excitement when a boy found one which made a cube. But only one group, out of seven, tried to collect the arrangements which made cubes together.

PATTERNS. The children were drawing patterns with rulers and compasses. One girl showed her pattern to her friends.

—Oh that's pretty. Did you make it up?

—Yes!

—True?

—Honest. I haven't anything to copy it from, have I?

FRACTIONS. $2\frac{1}{2} \times 2\frac{1}{2}$? Two children describe to each other how they do this.

—2 times 2 is 4, two halves are 1, one half of 2 is 1, half of half is a quarter and I added up to get $6\frac{1}{4}$.

—I did it by changing it all into halves, 5 of them, so 25, half of, $12\frac{1}{2}$.

SETS. A={9, 18, 31}. 'That's not a set,' says a girl. The teacher asks if she can give some other things that are not sets.

—Well, if you mix things up and wrote '5, cow, ship'. Her neighbours comment:

—Well, you might have five cows on a ship.

—They are all words, they are all on the paper—that's the relation.

—But so is 9, 18, 31, because they are on that paper. The first girl was busy with various coloured pens. She wrote a 5 in pencil, a 5 in red, a 5 in blue and the word 'sheep' in blue. Conversation continued:

—That's not a set, those fives are all the same.

—No, they are in different colours.

—But it's all a set, because you wrote it.

—Anything I think of would be a set. 9, 18, 31 is a set; a set is anything I can think of on earth.

A BOY. The second form is working at areas of triangles with great patience. One boy is drawing the numerals 202 on a square grid. He is making a number plate in metal work, but he wants the number to be larger than the template offered him. 'So I thought of doubling it up by making squares and doubling them.' He then draws a sketch of the crossbow he wants to make sometime. He says he has a go-kart in his bedroom which only needs a piston to work. 'But we can't use the lathes yet, because some boys in the class fool around. I wish they wouldn't, I've got so much to do.' When home-work is set, he joins in the chorus of disapproval without looking up from his drawing.

WRITE-UPS. With 1N on the sort of individual exploration lessons they've been doing the write-up seems very natural. They write quite casually about what happened in the lesson. . . . With 2C already the freedom of expression seems to have been lost. 'It's hard to explain, Miss'. Writing what happened seems rather silly and there is a tendency to make maths notes. . . . 6th form: 'An essay? I hate writing essays. I took up maths, so I wouldn't have to.' 'How can you write about maths?'

GAMES. Playing 'four-in-a-line'. A boys' team had played a girls' team and lost, through a bad move by one of the boys. The boy concerned had been heavily criticized by his team mates. I (rather foolishly) said, 'Who made the mistake?' The boy admitted that he had made it, but he was obviously very upset about it and on the verge of tears.

Later the same day, a group of teachers watched the game being played on a Madison project film. The use of such competitive games to stimulate effort to learn was raised. One teacher thought the use of competition could be a bad thing as the criticism by one's class-mates aroused emotions and anxiety.

This seems a matter worth discussing further some time.

A GIRL. She sits quietly writing 'I love David' on a set-square with her pen. Later the words get smudged a little as she uses the set-square to underline ANS.

HOW YOU ADD. How do you add 14 and 11?

'1 and 4 makes 5; then 1 and 1 makes 2 . . . 10 and 10 make 20; 4 and 1 make 5 . . . 1 and 4 (the numbers have been written on the board and this girl is "cross-adding", i.e. adding the tens digit of one number to the units digit of the other) makes 5; then 1 and 1 make 2. . . . Sir, you can add the two top ones to make 2 and the two bottom ones to make 5. . . .'

How do you add 14 and 11. I felt that many children could have had interesting things to say about this. But how are they to know at this stage that the teacher means the question. That it is of genuine interest to know how you, Mary, add 14 and 11. Naturally enough, Mary takes it that teacher wants to know how 14 and 11 is added officially. And memories of this can become a confused rigmarole. 'You' in the sentence can mean people, generally. But it was meant to be taken literally—you meant *you*. How does one establish the *you*?

FIRST LESSON. Discussion with the class about 'what they could do in arithmetic'. One boy said he could add, times, take away and divide two numbers. He was asked to give two numbers which he could add together. His first number was two thousand four hundred and fifty. A girl was given the chalk and asked to write this number on the blackboard. She wrote 2000450. Hilarious laughter from the class. The chalk was passed to a boy. He wrote 2000, 400, and 50 in three rows as if for an addition sum. More hilarious laughter from the class.

This laughter at other pupils' mistakes annoyed me. Nothing was said about the matter. But I hope that there is something in our method of teaching which will discourage this.

A CONVERSATION. That was interesting. I was surprised to see Stuart taking part, he is rather dull and normally just sits and dreams. . . . We could have saved a lot of time if some instruction could have been given first. Why did you say nothing? Didn't that part take a long time?

. . . I wonder whether allowing children time to come to their own decisions and making it clear that their answers are always acceptable isn't educationally valuable?

Yes I agree but I don't see where all the work is leading. . . . Well, accepting that the activity is worthwhile in the sense of encouraging children to be imaginative and to have confidence in their own ideas, it does also contain important mathematical experience which is essential for later work.

PREPARATION. A worksheet had been prepared by C. on number patterns. A number of questions followed the number square, to be used by the teacher. After discussion P. was concerned that he had not been able to use this worksheet in the same way as C. and that he had not got so much from it. The following day P. seemed relieved to report that he now knew why he had not, as he thought, achieved much with this worksheet. He decided it was because he had not prepared the sheet himself.

WHAT IS A POINT? This first discussion about a point came rather by surprise. The class had made some punched cards on which they had recorded certain details about themselves. We had talked for some time about boys who came by train and also stayed to lunch and so on. The cards were used to illustrate the idea of intersection. When the intersection of 'boys who came by train' and 'boys who came by bus' was suggested, the notion of an empty set was discussed. It was during this that one boy likened it to a point.

What is a point?

—It is a mere flick of the pen . . . a dot . . . it could be round or square depending what shape pencil you had.

How big is it?

As small as you can make it . . . smaller than that . . . you can't measure it . . . really it is just an idea, something in the mind. . . .

A few days later the class were asked to look at a white mark on the board for a few seconds. They were told to close their eyes and to think hard about the point.

Has anyone anything to say?

—It's grown bigger . . . opening up into a lot of circles . . . mine's gone . . . looks like a cow . . . falling to pieces . . . ball-like sun-bright shining orange . . . like flowers in a field . . . top yellow bottom red. I've lost it now. . . .

Comments of this kind continued. More and more motion was reported—the children seemed deeply involved with their dot and were not at all interested in each others comments. S's cow amused nobody. There was no indication that any of the stories were made up and I must admit to some fear at the excitement some children were getting from all this—on one occasion, W was talking about the blackish-yellow background eating his dot. He became very excited as some sort of climax seemed imminent. I told him to open his eyes—he seemed relieved.

Later I asked the children to try and keep a white dot in mind without change of position, colour or shape. After about a minute of complete silence with closed eyes I asked them to open their eyes and report what happened. There had been changes though these were very much more restricted than in the previous situation. Only three boys said they had managed to keep their dot fixed.

using resources

The young teacher joining a flourishing mathematics department should have a very different experience from the traditional, often overwhelming, isolation of the probationary year. He may also find that there is an increasing network of other agencies which can help or hamper, according to the way he uses them. Teachers' centres, Institutes of Education, conferences and courses, professional associations, inspectors, college tutors—all conspire, it seems, to bring various pressures upon a teacher. But it may well be worse to work in areas where some of these are not available, and we think that they are services that can be used. Much depends on the proper attitude towards the educational structure outside schools. So many of the important reforms in recent years have stemmed from classrooms that we feel teachers can justifiably place themselves at the centre of the network.

Here is the college tutor come to discuss teaching practice arrangements. Very well; at the same time he can be asked to work with your class on those days that you would like to work with a colleague in a neighbouring primary school. And the tutor's students could come in for one afternoon to join the staff in setting up, say, some work in small groups for the unstreamed first forms. Here is the local adviser come to make some arrangements for an overseas visitor. Very well; at the same time he can be asked to join a group of fourth formers whom you want to send away somewhere to make a film. And you can request him to organize an afternoon in which teachers from various schools could come and work with children together. Here is the researcher with various favours to beg. Very well; he can leave you that tape-recorder after he has used it in your classroom. . . .

The teacher's responsibilities are not to his headmaster or employing authority. He is responsible *to* his pupils. In fulfilling this responsibility, he may expect and demand the fulfillment of those responsibilities that are due to him.

aphorisms and argument

One day the Hodja took a ladder, propped it against the wall of an orchard, climbed over and pulled the ladder after him. Unfortunately for him, the gardener had seen everything.

'What are you doing here?' he demanded.

'I am selling ladders,' said the Hodja.

'Does this look like the place to sell ladders?' asked the gardener.

'What a question!' exclaimed the Hodja. 'Do you think there is only one place to sell ladders in?'

□ Everyone is a mathematician though he may not know he is. The mind knows more than it knows it knows.
□ In developing and using the powers that he already has there is no other feedback than his own judgement and decision, though these may be affected by the conventions of social agreement.
□ In the same moment we conceive a form and we agree with each other. Living mathematics is communal; it dies under competition.
□ Without variety of experience people cannot see their actions as choices. Play perceives possibilities.
□ It is difficult to sustain the constant recognition of the limits of our own knowledge. Ignorance is not bliss but it must be vigilantly embraced to preserve proper reverence for the as-yet-unknown.
□ In general, mathematics cannot be received; it has to be enacted.
□ Commencing in the cradle, it is a symbolizing activity and carries an emotional charge that can easily explode.
□ It can become a mindless mind training and a refuge from reality or—unfortunately all too easily—a weapon of subjection and a focus for fear.
□ Things signified are more important than their signs.
□ Impatient with our perceptions we surrender love to premature unification.
□ Mathematics can also enlarge the limits of the possible.
□ Apparently a uniquely difficult subject to learn it can also be a uniquely rewarding one to teach.

'How can one attain wisdom?' the Hodja was once asked.

'Always listen attentively to what those who know tell you,' he replied, 'and if someone is listening to you, listen carefully to what you are saying.'

some information

Association of Teachers of Mathematics (A.T.M.) Concerned with the whole range of mathematical education from infant school to university. Holds various conferences and is particularly active in more than fifty local branches. Publishes journals, pamphlets, books and films. Central office: A.T.M., Market St. Chambers, Nelson, Lancs. Annual subscription: £3.00.

A.T.M. branch Addresses of secretaries are published regularly in the A.T.M. journal, *Mathematics Teaching*.

British Broadcasting Corporation (B.B.C.) Various school programmes with associated literature. B.B.C. Schools Television, Television Centre, Wood Lane, London W.12.

College of Education

Computer Education Bulletin of the British Computer Society, distributed by Project Technology, College of Education, Loughborough. Annual subscription: 50p.

Certificate of Secondary Education (C.S.E.) Regional Examination Board

Department of Education and Science (D.E.S.) Various courses throughout the country, advertised direct to schools. D.E.S., Curzon St., London W.1.

Educational Studies in Mathematics International quarterly journal published by Reidel, P.O. Box 17, Dordrecht, Holland. Annual subscription: Dfl. 80.

Educational suppliers See p. 232.

H.M. Stationer's Office Atlantic House, Holborn Viaduct, London E.C.1.

Institute of Mathematics and its Application (I.M.A.) Maitland House, Warren Square, Southend, Essex.

Institute of Education At local university.

International Study Group for Mathematical Learning Publishes quarterly journal, distributed by Gordon & Breach, 12 Bloomsbury Way, London W.C.1.

Local Education Authority (L.E.A.)

Madison Project Publishes books, tapes and films. Syracuse University, Syracuse 10, New York, U.S.A.

Mathematical Association (M.A.) For teachers of mathematics in schools and universities. Local branches in various parts of the country. Publishes quarterly journal and various reports. Central office: M.A., 150 Friar St., Reading. Annual subscription: £5.

Mathematical Pie Journal for students, published by Mathematical Pie, 100 Burman Rd., Solihull, Warks. Annual subscription: 50p.

Mathematical Spectrum Journal for students, published by Oxford University Press. Annual subscription: 30p.

Mathematics for the Majority Project School Council project for the less able. Guides published by Chatto & Windus. Continuation project developing further material particularly for mathematics of environment. Director, Mathematics for majority continuation project, 3 The Cloisters, Exeter, Devon.

Mathematics in Education and Industry (M.E.I.) Liaison scheme sponsored by the Mathematical Association.

Mathematics Teachers' Forum Newsletter, formerly of the Nuffield project, published by Fanfare Bureau, 516 Forest Rd., London E.17. Annual subscription: 60p.

Midlands Mathematics Experiment (M.M.E.) Secretary, Harold Malley School, Solihull, Warks. Textbooks published by Harrap.

National Council of Teachers of Mathematics (N.C.T.M.) National association of teachers in U.S.A. Publishes journals, pamphlets and books. Central office: N.C.T.M., 1201 Sixteenth St.N.W., Washington D.C., U.S.A.

Nuffield Mathematics Teaching Project For range 5 to 13 years. Books and workcards published by Chambers and Murray jointly.

National Foundation for Educational Research (N.F.E.R.) 79 Wimpole St., London W.1.

Primary Mathematics Tri-annual journal, published by Pergamon Press, Headington, Oxford.

School Mathematics Project (S.M.P.) Director, Westfield College, Hampstead, London N.W.3. Holds various courses and conferences. Textbooks published by Cambridge University Press.

Schools Council Miscellaneous publications—e.g. C.S.E. examination bulletins, nos. 2 and 7 on mathematics—from H.M. Stationer's Office or bookshops. Schools Council, 160 Great Portland St., London.

Scottish Mathematics Group Textbooks published by Blackie and Chambers jointly.

Teachers' Centre

Technical College

University

some books

A.T.M., *Mathematics Teaching*, nos. 1– , A.T.M., 1952– .
A.T.M., *Supplement*, nos. 1– , A.T.M., 1965– .
A.T.M., *Examinations and assessment*, A.T.M., 1968.
A.T.M., *Development of mathematical activity . . .*, A.T.M., 1966.
A.T.M., *Some lessons in mathematics*, Cambridge U.P., 1964.
A.T.M., *Notes on mathematics in primary schools*, Cambridge U.P., 1967.
A.T.M., *Mathematical reflections*, Cambridge U.P., 1970.
Berger, J., *A fortunate man*, Penguin, 1971.
Blishen, E. (ed.), *The school that I'd like*, Penguin, 1969.
Boole, M., *Collected works*, vols. 1–4, Daniel, 1931.
Bourbaki, N., *Eléments de mathématique*, Hermann, 1950.
Cage, J., *A year from monday*, Calder & Boyars, 1968.
Davis, R., *Discovery in mathematics*, Addison Wesley, 1964.
Davis, R., *Mathematics teaching*, J. of R. & D. in Ed., monograph 1, 1967.
Douglas, M., *Natural symbols*, Barrie & Rockcliffe, 1970.
Downing, C., *Tales of the Hodja*, Oxford U.P., 1964.
Fuller, B., *Education automation*, S. Illinois U.P., 1962.
Gattegno, C., *The adolescent and his self*, Ed. Explorers, 1963.
Gattegno, C., *For the teaching of mathematics*, vols. 1–3, Ed. Exp., 1964.
Gattegno, C., *Mathematics with nos. in colour*, bks. 1–6, Ed. Exp., 1964.
Gattegno, C., *Towards a visual culture*, Outerbridge, 1964.
Gattegno, C., *What we owe children*, Kegan Paul, 1971.
Goutard, M., *Mathematics and children*, Ed. Explorers, 1964.
Holt, J., *How children fail*, Penguin, 1969.
Hourd, M., *Emotional aspects of learning*, Heinemann, 1956.
Jones, R. M., *Fantasy and feeling in education*, London U.P., 1970.
Laing, R., *The politics of experience*, Penguin, 1967.
McLuhan, M., *The gutenberg galaxy*, Kegan Paul, 1962.
Meredith, P., *Instruments of communication*, Pergamon, 1966.
Papy, G., *Modern mathematics*, vols. 1–2, Macmillan, 1968.
Piaget, J., *Moral judgement of the child*, Kegan Paul, 1950.
Polanyi, M., *Personal knowledge*, Kegan Paul, 1958.
Polya, G., *Mathematics and plausible reasoning*, Princeton, 1954.
Renwick, E. M., *Children learning mathematics*, Stockwell, 1963.
Rogers, C., *On becoming a person*, Houghton Mifflin, 1961.
Shah, I., *Exploits of Mulla Nasreddin*, Cape, 1966.
Spencer-Brown, G., *Laws of form*, Allen & Unwin, 1969.
Steiner, G., *Language and silence*, Faber, 1967.
Watzlavick, P., et al., *Pragmatics of human communication*, Faber, 1968.
Wilder, R. L., *Evolution of mathematical concepts*, Wiley, 1968.
Winnicot, D. W., *Playing and reality*, Tavistock, 1971.
Wittgenstein, L., *Foundations of mathematics*, Blackwell, 1956.

II Starting points

II Starting points

introduction

In this section, we give a miscellaneous selection of mathematical situations which may be used as starting points for further investigation. Most of these could be started with classes of any age or ability and investigated until appropriate conclusions have been reached.

To be most fruitful, such situations should be able to be developed in many directions. They must be able to be initiated simply, immediately, and with a minimum demand upon existing vocabulary or technique. For there to be any creative response, they must be presented with flexible and experimental intent, often with deliberate ambiguity.

We would emphasize that working from situations in the way we have suggested is not the same as working from problems. Part of the activity is, in fact, the formulation of problems that may arise out of the definitions and rules that are developed in discussion of the situation. Students will readily wish to solve problems that they have created themselves and such solutions are part of the work that follows from the starting point.

It should also be recognized that there is no single, correct, way of mathematising a situation. In the first instance, there is a tentative and roundabout process of making hypotheses, finding analogies, and seeking social agreement. At any particular time and place, this will depend on the people involved. At a later, and possibly more formal, stage, there will be attempts to condense the previous activity in order, for example, to solve problems. Symbols may be created and manipulated to assist the memory, or to facilitate the process of becoming operational. Any final record of the work will depend on the particular structures and symbols that were called upon, or created, during the activity. And 'sufficient unto the day is the rigour thereof.'

In selecting, or recognizing, starting points that will be fruitful, the teacher may wish to seek a balance between those that are genuinely open for himself as well as his students, and those that he knows through experience or insight are very likely to lead to structures already known to him if not to his students. In the latter case, it may be difficult for him to restrain his influence on the choices that have to be made. We do not suggest that he should not make choices, but think it is important that he should not only know when this is happening but be able to show it.

Will someone mark six points on the blackboard?

Anyone is able to take the piece of chalk and do this.

(Though they may not wish to. We observe, incidentally, that many children do like using the blackboard during lessons and we think it is useful to consider the blackboard as a communal recording tablet, available to anyone as discussion proceeds.)

With six points on the blackboard we are immediately in a situation that is, in fact, rich in possibilities. But how can they be realised? What happens next?

To begin with, the teacher must surely have a genuine interest in the response. How will this particular person arrange his points? Will he make a free choice, or will he be trying to supply what he thinks is required? Will he be mystified, wondering what game the teacher is playing, or will he be a zestful partner in an exploration.

What do teachers say when children ask them what the point is of doing some particular work? We believe the teacher ought to be able to say, 'Because I asked you'. But this presupposes an atmosphere of mutual commitment and trust, and perhaps in such an atmosphere the question is not asked. The teacher has to earn the right, it seems, to demand a creative and willing response to an imposed task, by sedulously taking care that he commits himself at times to the children's choices. So our account of developments from some starting points is misleading if they are interpreted as prescriptive paths to be followed. We have recorded particular choices that have been made at particular times. These can only be the shadows of other choices at other times.

Perhaps, to return to the six points, on one occasion others come out to the board to make further arrangements. Perhaps people are invited to talk about these arrangements. Some may be thought to be the same. In what sense can they be the same?

Now mathematics has inevitably commenced when it is decided how things are to be treated as the same. Among other issues, this raises the matter of how many different things there are. Equivalence is a choice at our disposal; but when a choice is made, counting must yield a unique answer. Here is the essence of the activity. We are free to choose bounds and may then explore the inexorable implications of our choice. But the activity must include the choice as well as the exploration.

We have based one starting point (p. 139) on an occasion when the discussion of equivalences of arrangements of six points was clarified by joining points to each other by straight lines. This led to interesting configuration problems. We have quoted an account of this work on p. 44.

Another activity based on six points is described on p. 82. This arose when, at the suggestion of some children, joins were made by lines that were not necessarily straight. In order to make every possible join, some crossings became inevitable. Immediately a search began for ways of joining with least crossings. In this case the underlying (topological) equivalence remained implicit.

We have discussed a particular example to indicate the sort of background to our starting points. We have not attempted a comprehensive selection and there are many other situations which would have been useful. We do not present the material in any uniform style. In some cases, we suggest various ways in which work could develop over a period; others we leave as isolated topics. Sometimes we mention further issues so that the reader may see how the work relates to other developments. None of the starting points have any particular age or ability in mind. Most of them can be, and have been, investigated in some way with various classes. Inevitably, we have borrowed ideas from colleagues and from books; but we have deliberately not given further references in this section. It is the process and not the product with which we are here concerned.

Our accounts imply some preliminary group discussion, with probable development into individual or small group work. Where it is helpful to have a planned sequence of interventions in mind, suggestions have been given. These will surely be translated into appropriate forms; they are merely a sort of safety-net, neutral but—we hope—reliable. Some indication of likely responses has also been given. But such is the originality and sophistication of comments made by children in these situations that no two lessons will be alike.

The role of the teacher seems to us to be quite crucial. Briefly we would recommend an attitude of watchful waitingness. In the teaching of mathematics, haste is unchaste.

area on a square lattice

Materials required: 5×5 geo-boards and some elastic bands. (See p. 200).

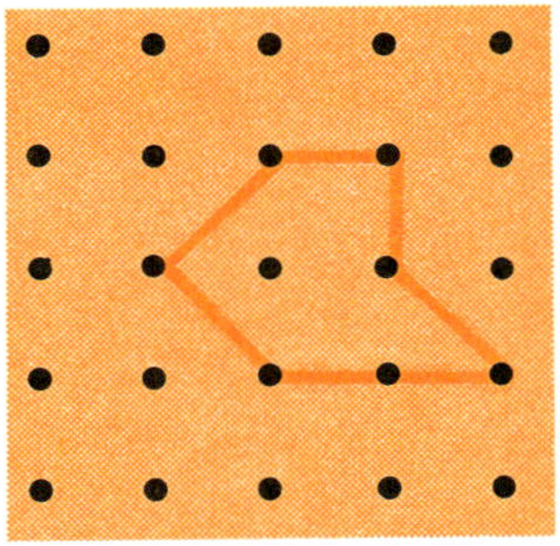

Make a shape using 8 nails. The 'using 8 nails' is to include those touching the band and those inside. What can you say about the shape you have made?

The replies to this will vary according to the interest of the pupil.

—It has 6 sides.
—There are 7 nails touching and 1 inside.
—It is $3\frac{1}{2}$ squares large.
—It is seven eighths of a square.
—. . .

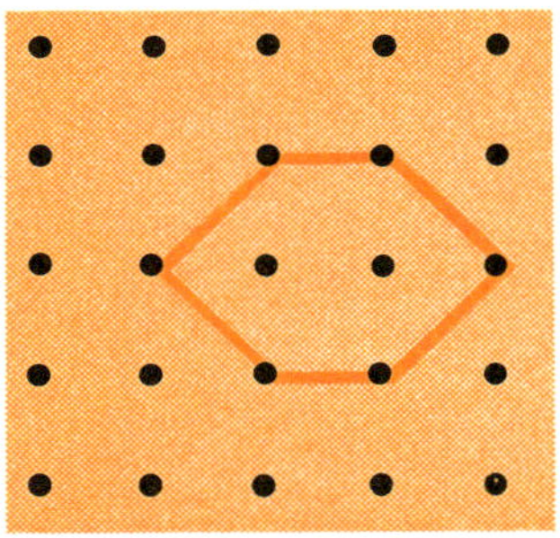

Make another shape using 8 nails.

This time certain parts of the information will have changed whilst some of it remains the same.

What is its area?
How many nails touching? Inside?
How many sides?
— . . .

Sometimes children notice the number of nails at vertices, the number on edges and the number inside.

corners	6	6	
edges	1	0	
inside	1	2	

Make as many '8 nail' shapes as you can.
What differences can you find?
What similarities?

Some get involved in all the different shapes they can make with 8 outside and 0 inside, 7 outside and 1 inside, . . . Others become occupied with areas and nail positions.

nails touching (T)	8	6	4	7
nails inside (I)	0	2	4	1
area (A)	3	4	5	$3\frac{1}{2}$

—Look! The area changes as the number of nails inside changes.

What is the smallest area? When is this?
What is the largest area? When is this?
What are the actual values of T and I in these cases?

Consider a different number of nails, say 6, 10, or 11.

How do the areas vary now in relation to nails touching and nails inside?
What is the largest area for a given number of nails?
The smallest area?
Can you find a relation between T, I and A?
(Pick's theorem: $A=\frac{1}{2}T+I-1$, cf. p. 207).

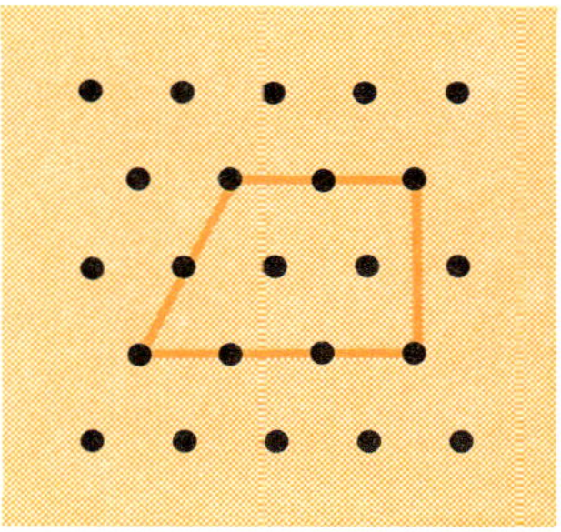

This theorem relates area to the number of nails. Is it possible to relate perimeters in this way?

Further investigations could be carried out on triangular or hexagonal lattices.

arranging five counters

Materials required: 5 cardboard counters (black on one side, white on the other) for each pupil. Some sticky gummed circles (or squares)—of the same colours. Strips of white paper, about 20 × 5 cm.

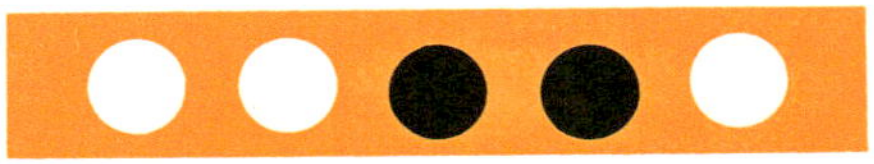

Arrange your five counters in a line across the desk.

It is interesting to note initial choices to see if any arrangement predominates—is symmetry preferred?

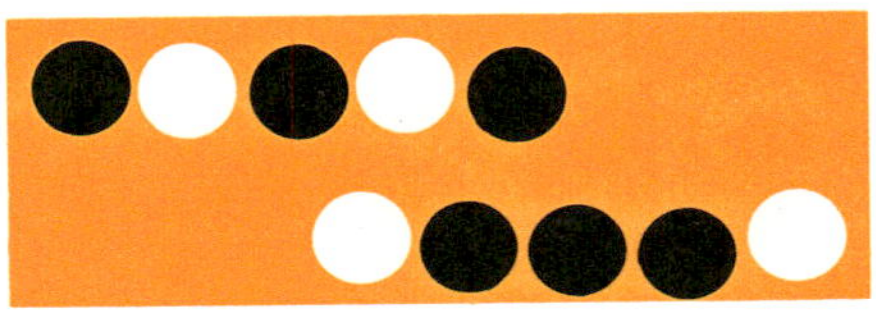

Copy your arrangement by sticking your gummed shapes on to the strip of paper.

Pin them on the board.
How many arrangements are there?

At this stage pupils begin to talk about arrangements that are the same and this idea needs to be clarified—do they mean identical, has the same number of blacks as, . . . ?

Remove duplicates from the board.
Are there any other arrangements?
Make them, stick them on the board—but no duplicates.

Are there any more?
How do you know?

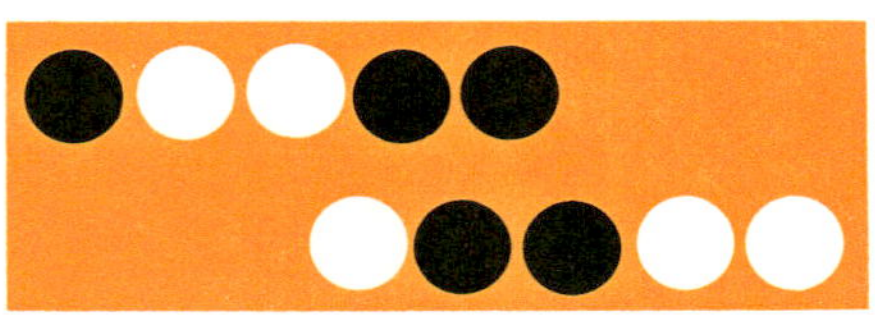

There will be suggestions of why there are no more as pupils begin to classify them in certain ways.

For example: (a) Opposites BBBBB and WWWWW
BWWBB and WBBWW

Does every arrangement have an opposite?
Can an arrangement be its own opposite?

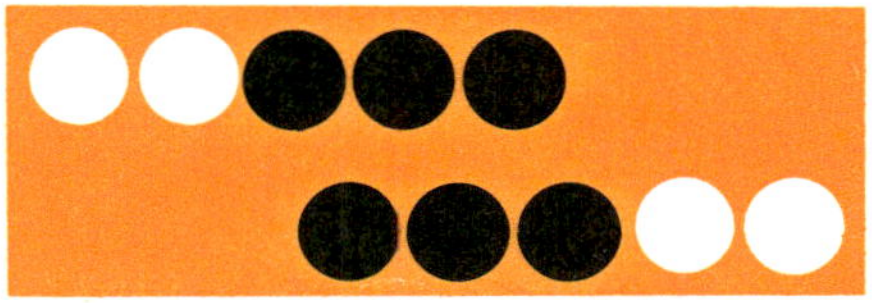

(b) Reverses WWBBB and BBBWW
BBWBW and WBWBB

Does every arrangement have a reverse?
Which ones do not? Why not?

How many pairs? How many singles?

Are there any connections between opposites and reverses?
Can a pair be reverses and opposite?

(c) All those with no blacks, one black, two blacks. . . .

How many sets will there be?
How many in each set?

Can you see why some sets have the same number of arrangements?
Where are the opposites and reverses now?

The activity of classifying according to different criteria is useful in its own right, but if the problem is investigated for different numbers of counters then Pascal's triangle will emerge.

Investigate the numbers of arrangements for 1, 2, 3, 4 similar counters and classify in some way.

bracelets

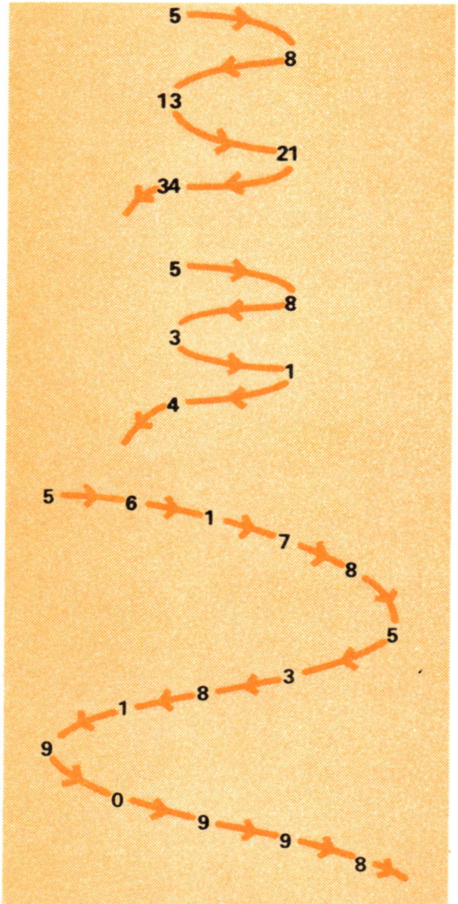

A sequence of numbers is linked by arrows on the blackboard.

Do you notice anything about this sequence?

What is the next term? And the next?

Pupils may or may not recognize the Fibonacci sequence where each term after the second is the sum of the previous two. They could be invited to find the rule or they can be told it.

After this preliminary activity another sequence is written on the board.

What is the next term? And the next?

These numbers are the units digits of the first sequence. They could also be generated directly with all sums being taken mod 10, e.g. the third term is $5+8=3 \pmod{10}$. It does not matter how the rule is stated as long as it can be used.

Choose two digits and start a chain yourself using the last rule.

Does it go on for ever?

Someone will notice that his chain is repeating and goes back to the first number to make a cycle of numbers. One class, in fact, called the cycle of numbers a 'bracelet'. Different starting points may produce the same bracelet.

There are 5 different bracelets—the largest of these has 60 digits in it.

These bracelets can be collected and explored further.

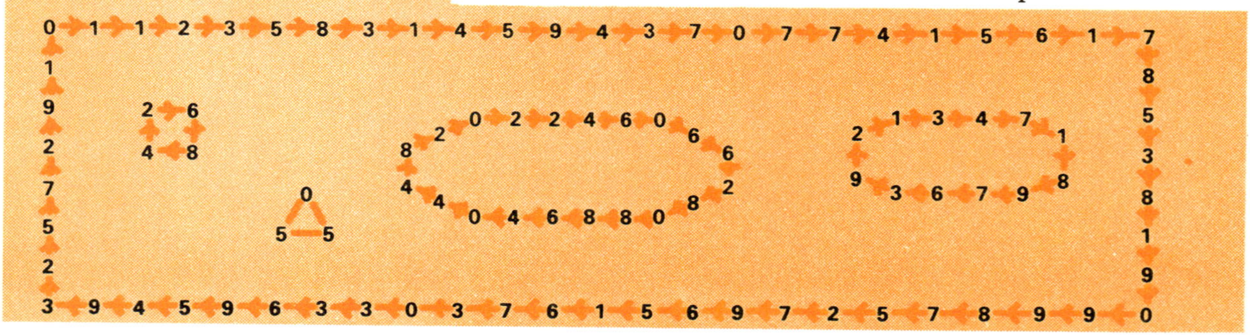

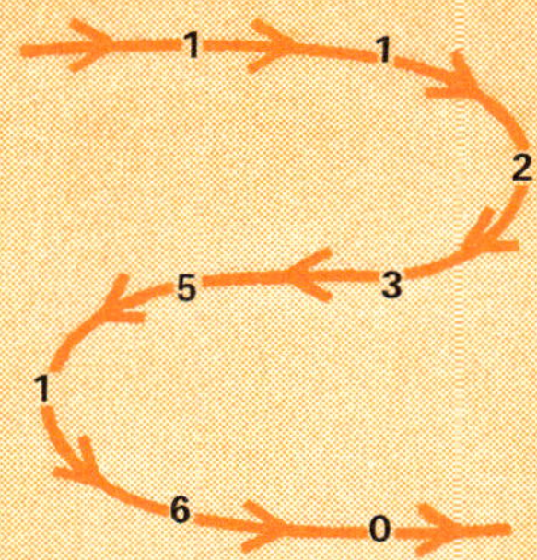

Alternatively the situation may be generalized further before systematic investigation begins. Another sequence is put on the board, using the same additive rule but this time sums are taken, say, to mod 7. The rule may be seen as adding numbers in base 7 and retaining only the units. Once again it does not matter how the rule is stated as long as it can be used. It is often helpful to use chalk of a different colour when switching to another modulus.

Are there bracelets this time?

The modulus may be changed again. The various bracelets that arise in each case are recorded and a systematic investigation encouraged.

How many bracelets are there in each case?
How big are they?
What patterns can be found in each bracelet?
Where are the zeros?

Some further issues:

1. Which terms of the Fibonacci sequence, 1, 1, 2, 3, 5, . . ., are divisible by 5?
2. Investigate the size of the various bracelets when the modulus is prime. Can you find a pattern?
3. The terms of the usual Fibonacci sequence may be derived by calculating successive powers of a certain matrix A, as shown. The bracelets problem may be investigated by considering powers of A, mod n. For what values of k will $A^k = A$?

   The problem may be generalized to any 2×2 matrix A in various finite arithmetics. For what value of k will $A^k = A$? Is k the same for any matrix?

$$A = \begin{pmatrix} 0 & 1 \\ 1 & 1 \end{pmatrix}$$

$$A^2 = \begin{pmatrix} 1 & 1 \\ 1 & 2 \end{pmatrix} \qquad A^3 = \begin{pmatrix} 1 & 2 \\ 2 & 3 \end{pmatrix}$$

$$A^n = \begin{pmatrix} f_{n-1} & f_n \\ f_n & f_{n+1} \end{pmatrix}$$

4. For the sequence used in the starting point the n-th term u_n is the sum, mod 10, of the two previous terms. This may be expressed as $u_n = u_{n-1} \star u_{n-2}$ where the operation $\star$ is here addition, mod 10. This was then changed to addition, mod 7, and so on. Now consider some other operations and the corresponding bracelets. For example, with *multiplication*, mod 7, there is a bracelet of 3 terms, one of 8 terms, and one of 23 terms.

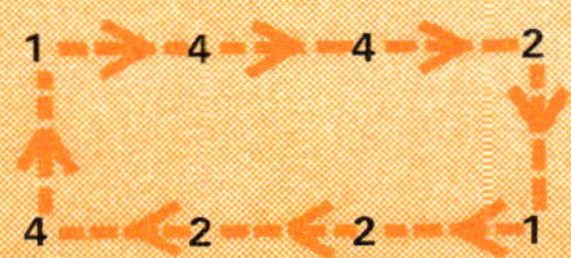

choosing rods

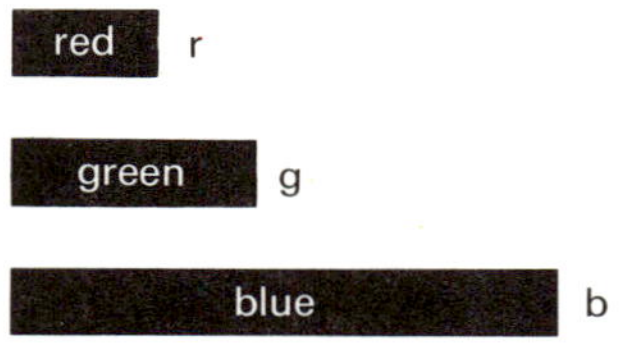

Place at least 30 coloured rods (e.g. Cuisenaire—see p. 186) in a box or bag using only 3 colours, say, red, green and blue, with at least ten of each colour. Offer the box to a pupil.

Take any 3 rods from the box.
Move on to another person.
Take any 3 rods.
Move on again.
Take 3 rods . . .

This is continued until, say, 10 people have 3 rods each. These people can be asked to hold up their rods.
What colour rods have you?
Has anyone got three of the same colour?
What colour are yours?—any the same?

r^3	g^3	b^3
r^2g	g^2r	b^2r
r^2b	g^2b	b^2g
rgb		

As the colours are given these could be recorded on the board and some abbreviated notation introduced, e.g. (blue, red, red) becoming (b r^2) or (brr). Order is usually ignored; when differences in order are raised discussion tends to claim that such triples are nevertheless the same!

How many 'different' sets of rods have we?
Are there more?

This could be followed by further choice of rods or use of the notation.

How many different sets are there?
3 rods chosen from 3 colours produces 10 combinations. How can we be sure we have all the possibilities?

r^2	g^2	b^2	y^2
rg	gb	by	
rb	gy		
ry			

Some rods are now put in the box using 4 colours.
Choose any 2 rods. Move on to another person. *Choose any 2 rods. . .*
How many different sets of 2 rods have you?
How many are there?
This again produces 10 different sets.

number of rods	number of colours	number of ways
2	1	1
2	2	3
2	3	6
⋮	⋮	⋮
3	1	1
3	2	4
3	3	10
⋮	⋮	⋮
4	1	1
4	2	5
⋮	⋮	⋮

Make a table showing the numbers for rods taken, colours, and different ways.

Use rods or the notation to find other numbers of sets for a given number of rods to be chosen from a given number of colours.

It may be useful to have the class working in groups to produce as many results as possible. These can be recorded on the blackboard and discussion of the results can follow. It may seem trivial to deal with the case of only choosing one rod.

What do you notice about the 'number of ways' when 2 rods were chosen? When 3 rods were chosen. . . .

This may be a question of seeing the pattern of the number sequence or a recognition of its place in Pascal's triangle.

number of rods chosen ↓ \ number of colours →	1	2	3	4	5	6	7
1	1	2	3	4	5		
2	1	3	6	10	15		
3	1	4	10	20	35		
4	1	5	15	35	70		
5	1	6	21	56	126		
6							
7							

Make a table like the one shown and continue it.

Where it has been met before the table will recall Pascal's triangle. Otherwise inspection may yield the observation that each number is the sum of the numbers immediately above and the one to the left. It is usually swiftly induced that this additive structure holds throughout. But how confident can we be that this is so? Are the numbers throughout the ones from Pascal's triangle? Further considerations may or may not be appropriate in different cases.

The choice of 3 rods from a set of rods in 3 colours may be indicated by a vector the colours being in some agreed order, e.g. (3, 0, 0) for r^3, (2, 0, 1) for r^2b and so on. The sum of the elements of such vectors is constant. How many vectors are there?

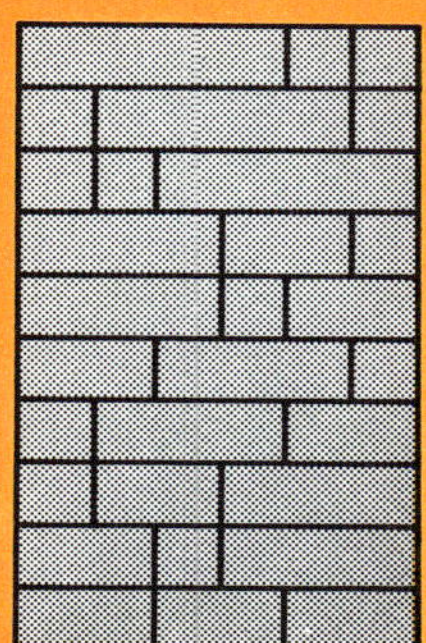

Make these partitions with Cuisenaire rods.

Note that in each row of the 'wall' there are joins of bricks in 2 out of 5 possible positions. What is the number of ways of choosing 2 out of 5?

In the corresponding wall for the choices of m rods from a set of rods in n colours how many joins are there in each row? In how many possible positions?

co-ordinates

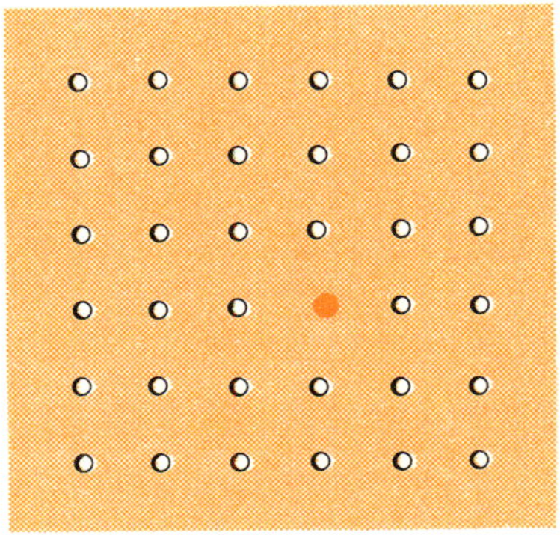

Materials required: a nail board (6 × 6 or larger), and something that will fit over the top of the nails (golf tee heads are useful—you need at least one of one colour and one of another). Alternatively a peg board and pegs would suffice but it would be more convenient if the holes in the peg board are 3 to 5 cm apart.

A coloured peg is placed in one of the holes.

Can you give a name to this peg?

Replies seem to vary from 'Tom' to 'Mary' to 'Green' to 'One' . . .

The peg is moved to another hole.

Can you give a name to this peg?

Describe where it is.

Move the peg again, again, again. . . .

The 'name' is now likely to be 'in from the left, up from the bottom' or 'down from the top and in from the right' or . . .

These descriptions are gradually refined and a class agreement can be reached which only requires a number pair. The convention of using the bottom left hand hole as (0, 0) can be elicited from a pupil and given his name.

Fred says that he will refer to a position for the peg as so many holes in from the left hand side and so many holes up from the bottom. So when we say (3, 4), by Fred's rule, we mean along from left 3 and up from bottom 4.

It is, of course, not necessary to have the origin in one particular place but useful to extend Fred's rule to mean just 'so many to the right and so many up'.

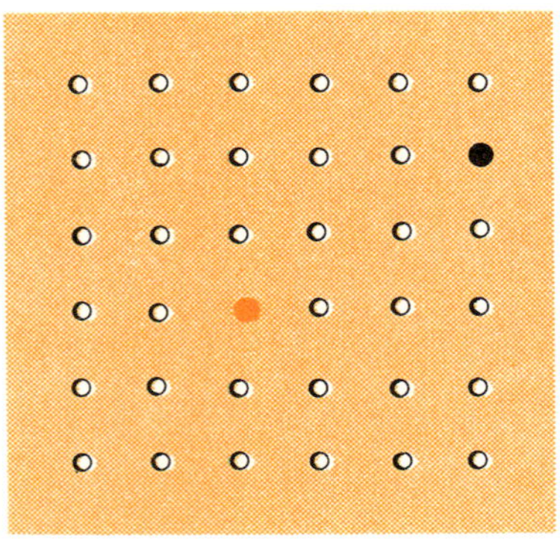

The coloured peg is placed near the middle of the board.

I will call this peg (1, 2).

A black peg is placed on the board.

What is the name of the black peg?

It is moved to different positions.

What is the name of the black peg now? . . . and now?

At this stage it seems wise to avoid negative numbers but to include the (0, 3) and (3, 0) type of situation.

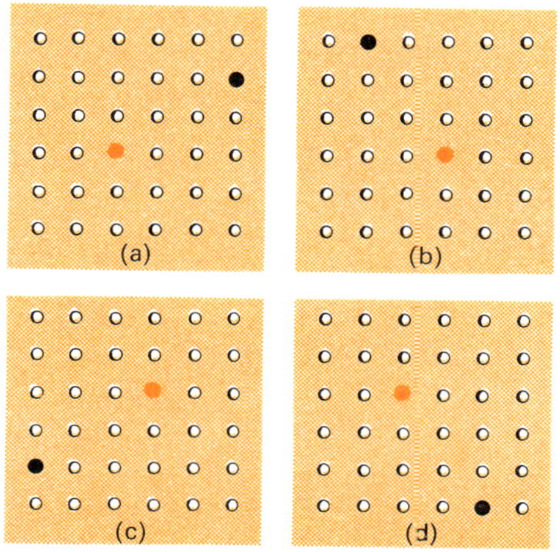

With the coloured peg near the middle of the board.
I will call this peg (4, 5).
A black peg is placed as shown in fig. a.
What is the name of the black peg?
What will its name be if the board is rotated through 90° anti-clockwise? (fig. *b*). *The coloured peg retains the name* (4, 5).
What happens if it is rotated a further 90° anti-clockwise? (fig. c). And again? (fig. d). And again?
The coloured peg can be given new names and the black peg can be moved into new positions.
Can you say the names of the black peg before the board is turned?

In the early stages the position of the black peg, relative to that of the coloured peg, should be chosen in such a way as to avoid negative numbers after rotation. At a later stage it is interesting to place the black peg in positions which do require negative numbers after rotation. 9-year-olds have been known to invent negative numbers to cope with this situation.

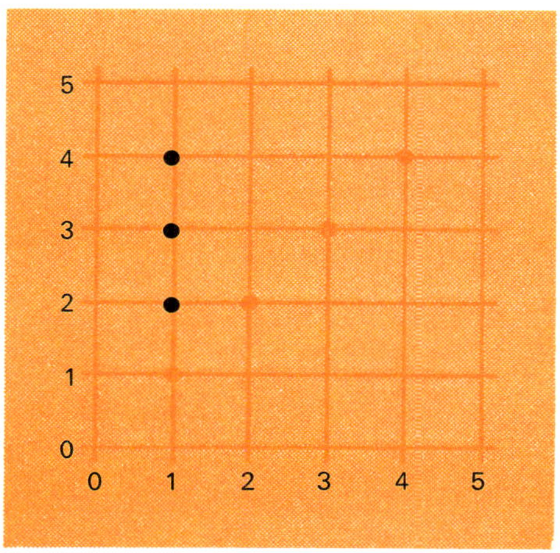

Four-in-a-line

If the activity just described is followed by this game then consolidation of the convention is possible alongside the excitement of game strategy.

The game of four-in-a-line is played as follows:

There are two teams (or two people).
Each says a pair of numbers in turn and a 'marker' puts the point on the grid (the teams must not touch the grid). The aim of the game is for one of the teams to obtain four points in a line (the lines can be horizontal, vertical or diagonal).

counting-out

Do you know any 'dipping' rhymes?
A collection is made on the blackboard.
—Ibble obble, black bobble
Ibble obble out.
—One potato, two potato, three potato, four
Five potato, six potato, seven potato, more.
—Your shoes need cleaning.

Six pupils are seated in a row facing the class, and numbered 1 to 6. Someone is invited to come and 'dip' using the first rhyme. The pupils are standing at the beginning of the dipping and sit as they are counted out. The winner is the last one standing.

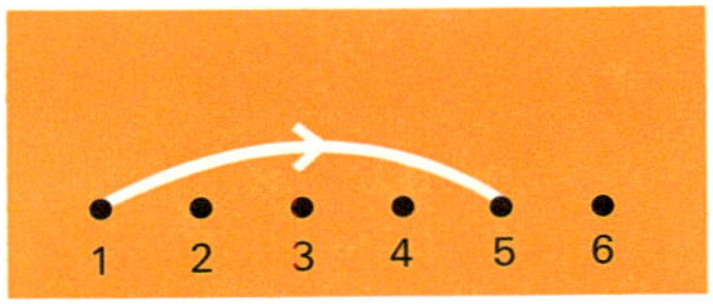

With which number did you start?
Which number won?
This is recorded on the board.

Do you think the same person would have won if one of the other rhymes had been used?

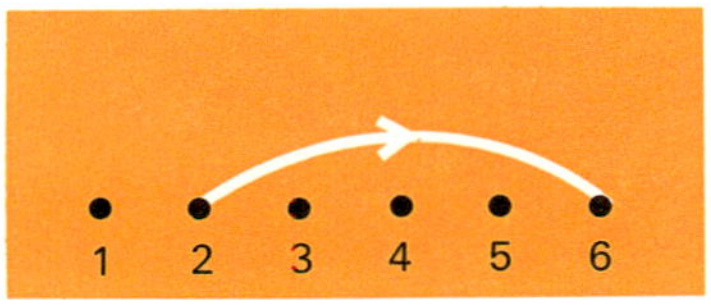

If we use the same rhyme but start dipping at number 2 who do you think will win?
The dipping is performed. Number 6 wins.

—It's always four on.

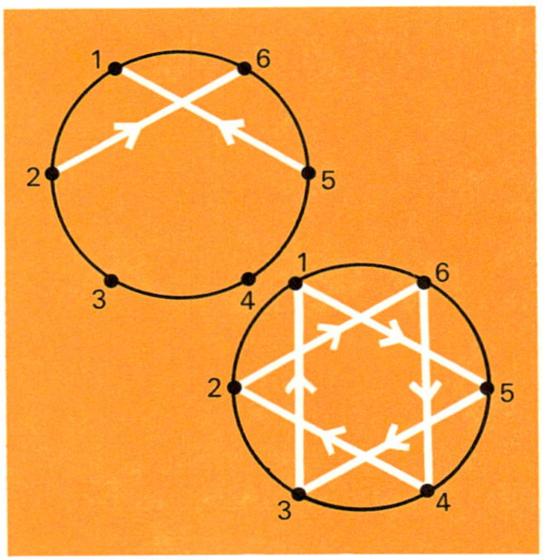

Which number would win if we started with number 4?
—Number 2, because it's four on and it is like counting round a circle.

The six dots might be redrawn in a circle. It is checked that when starting with number 4, number 2 wins.

Could someone finish the graph?

Perhaps we should test one other starting number to make sure the graph is correct. . . .

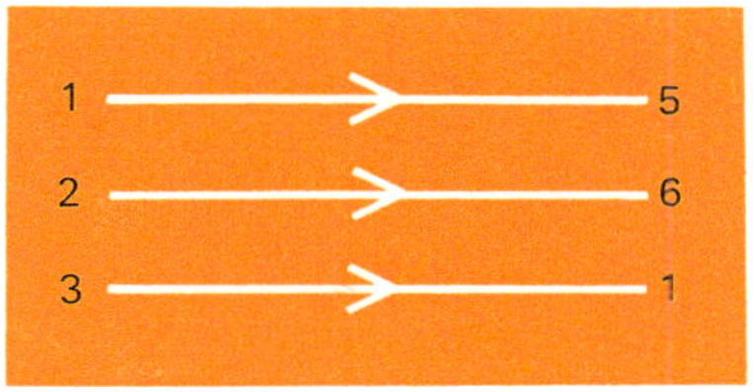

The results might be written in a table and the modular pattern observed: winning number = starting number + 4 (mod 6).

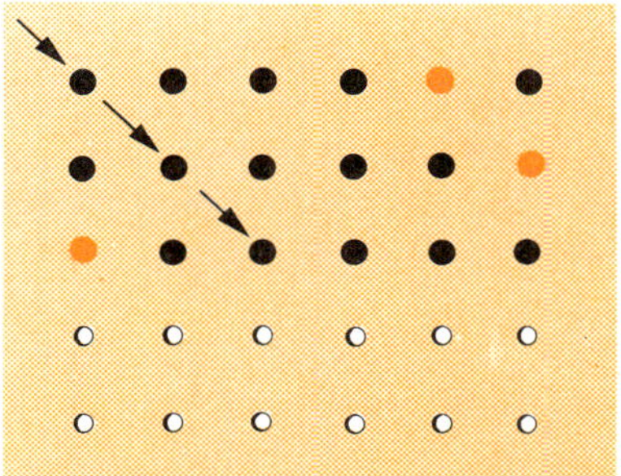

There are several possibilities for further investigation. Results of these might be recorded in table form; or on duplicated sheets of circles with various numbers of points on the circumference; or another useful medium is pegboard and pegs. The figure shows three counts starting at different pegs. The arrow indicates the starting peg and the winning peg is shown coloured.

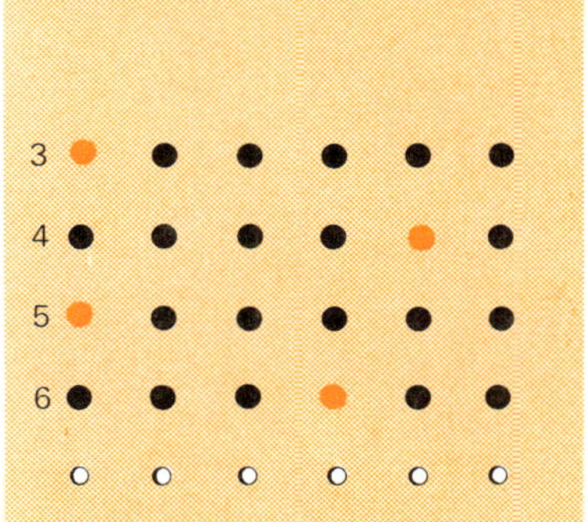

One group can investigate what happens as they vary the number of stresses in the rhyme, keeping the number of people constant. The numbers in the figure indicate the number of stresses in each case.

Another group can vary the number of people but keep the number of stresses in the rhyme constant.

Another can look at the order in which numbers are counted out. For example, with five people numbered 1, 2, 3, 4, 5, counting out in twos, the order of being counted out is 2, 4, 1, 5, 3. When counting out in threes, it is 3, 1, 5, 2, 4; and in fours, it is 4, 3, 5, 2, 1.

Other problems and issues will arise.

crossings

Six points are marked on the blackboard.

Join any two of the points.
Join any other two points.

. . . repeated until all joins are made. It may happen that questions are raised whether joins may cross, whether joins must be straight, and so on. Decision on such issues is left to each individual. The teacher merely insists that every possible join be made.

How many crossings are there?

There will always be some—here, 9.

Could we have done it with less?

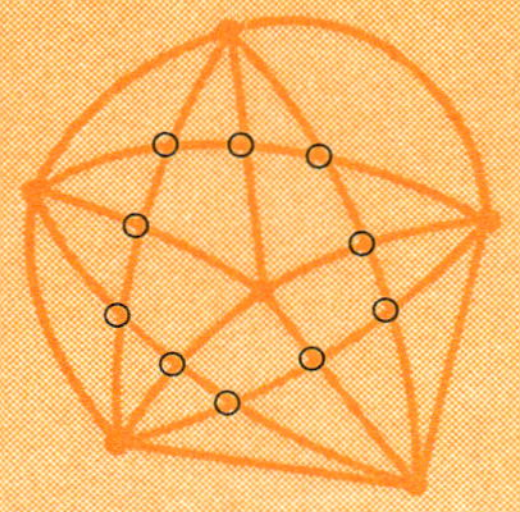

The board is cleared and the construction attempted with six more points. This will invariably yield only 3 crossings, with some conviction—possibly after further attempts—that it could not be done in less. Some systematic ways of making joins may emerge.

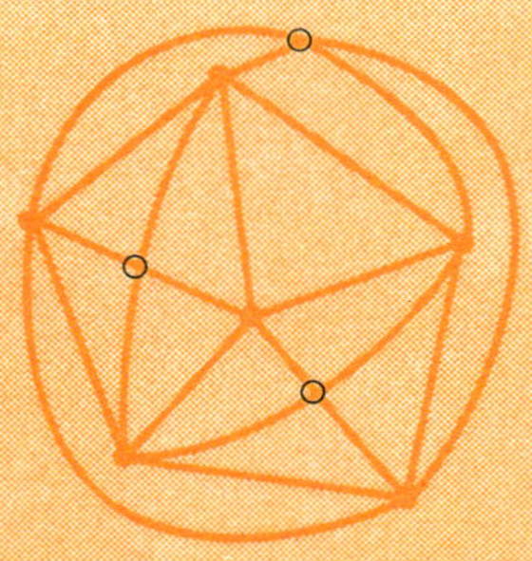

What if we started with 8 points?

There may be some guesses forthcoming. Some students will start doodling. . . .

Any resulting claim for the least number of crossings may be passed round or exhibited on the blackboard. The degree of conviction that the minimum has been achieved may not be as unanimous as in the first case.

number of points	least number of crossings
.	.
4	0
5	1
6	3
.	.

What happens with other numbers of points?
Are there any definite ways of getting least crossings?
Is there any pattern in the results?

(A warning—there are some very difficult and still unsolved problems here. The approach to these will vary in different cases but there will always be considerable interest in the making and refuting of conjectures.)

Some further issues:

1. A diagram of points and joins with no crossings is called a *planar graph*. With more than four points such a diagram will not include all possible joins. What joins there are will form well-defined regions. Is there any relation between the number of points, the number of joins and the number of regions?

points	5	5	.
joins	6	4	.
regions	3	2	.

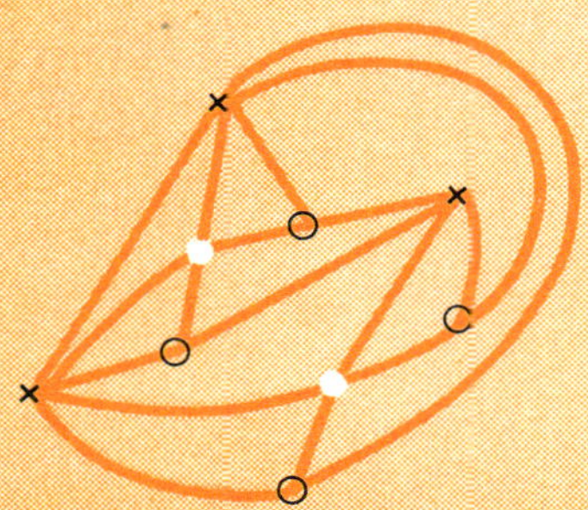

2. Mark 3 points with a cross and 4 points with a circle. Each cross must be joined by a line to each circle.
   What is the least number of crossings with which this can be done?
   Generalize the situation.
   (The case of 3 crosses and 3 circles is sometimes known as the three utilities problem: three houses have to be connected to gas, electricity and water supplies. . . .)

calculating machines

This assumes the ability to use a desk calculator for addition, subtraction, multiplication and division. (See p. 172).

The first stages of this activity are simply to establish the operation of subtraction.

Use your machine to take 5 from 8; 5 from 7; 5 from 6 . . . 5 from 5; 5 from 4. What do you notice?

Take 3 from 1; 7 from 5; 12 from 10. What do you notice?

Take 5 from 0; 3 from 0; 1 from 0. What do you notice?

At this stage the continual ringing of bells and occurrence of 9's will show some pattern. The 5 from 0 produces . . . 99995; 3 from 0 produces . . . 99997; 1 from 0 produces . . . 99999.

Find out what other numbers are produced when 2, 4, 6, 7, . . . are taken from 0.

Subtracting numbers from 0 produce the long numbers with 9's. To avoid having to say the whole number a shorthand could be used, e.g. . . . 99997 (0 − 3) could be called *twiddle* 3 and written (∼3). (∼4) would be . . . 99996 and so on.

Can you see a quick way of getting the twiddle number? What would it be for:

∼7? *∼9?* *∼15?* *∼38?*

How would you write the *twiddle* numbers

99992? *99990?* *99983?* *99957?*

13 + (∼7) = 6
(∼7) + 13 = 6
13 − (∼7) = 20
(∼7) − 13 = ∼20

Various problems can now be suggested, e.g. adding ordinary and twiddle numbers; subtracting them; multiplying and dividing them; taking squares and square roots. . . .

distribution of sexes

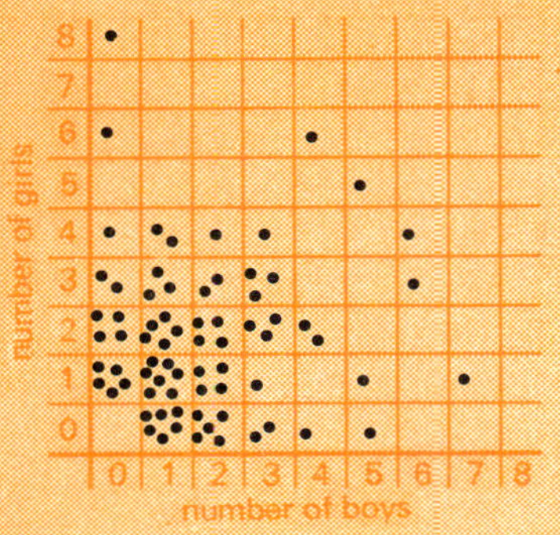

number of girls \ number of boys	0	1	2	3	4	5	6	7	8
8	1	0	0	0	0	0	0	0	0
7	0	0	0	0	0	0	0	0	0
6	1	0	0	0	1	0	0	0	0
5	0	0	0	0	0	1	0	0	0
4	1	2	1	1	0	0	1	0	0
3	2	3	2	3	0	0	1	0	0
2	4	5	4	3	2	0	0	0	0
1	5	7	4	1	0	1	0	1	0
0	0	6	5	2	1	1	0	0	0

1 1

1 2 1

1 3 3 1

1 4 6 4 1

1 5 10 10 5 1

How many children are there in your family?
How many boys? How many girls?
(Don't forget yourself).

A discussion of how this can be recorded for a large group could lead to placing points in squares as shown, to produce a scattergram.

What do you notice about the scattergram?

Colour the squares where the number of boys is the same as the number of girls.

Are there more points above the line of squares or below the line of squares? Comment!

To investigate these results further it is useful to record the information as a matrix.

Colour the squares showing families with 3 children. How many are there? What proportion of all the families?

How many families of 2 children? . . . 4 children?

What is the most common size of family?

At this stage information could be collected from another class and another matrix formed. To give a larger sample the two matrices could be added.

Look again at the families of 4 children.

How many are all boys? All girls? 3 boys and 1 girl? 3 girls and 1 boy? 2 boys and 2 girls?

What have your results to do with the row of Pascal's triangle?

Compare other family sizes with rows of Pascal's triangle?

Make a stereogram of your matrix (3-dimensional graph built up on squared paper with cubes, see p. 184.)

end digits

What can you say about the result of $5 \times 5 \times 5 \times 5 \times 5 \times 5$?

—It's large.
—It's 5 to the power of 6.
—Ends in 0.
—It ends in 5.

What digit does it end in?
(There is usually disagreement over this question).
Has anyone calculated the result?
What digit does it end in?
Do we need to calculate the result to find the end digit?

What is the end digit of $5 \times 5 \times 5 \times 5 \times 5 \times 5 \times 5$?

5^1	ends in	5
5^2	ends in	5
5^3	ends in	5
5^4	ends in	5
5^5	ends in	5
5^6	ends in	5

A table is built up on the blackboard.

How about 5^{10}?
(Many pupils think the end digit of this number is 0).
How about the end digit of 5^{99}?

Can you find other numbers which behave in this way?
It is found that of other numbers less than 10 only 1 and 6 behave in a similar way.

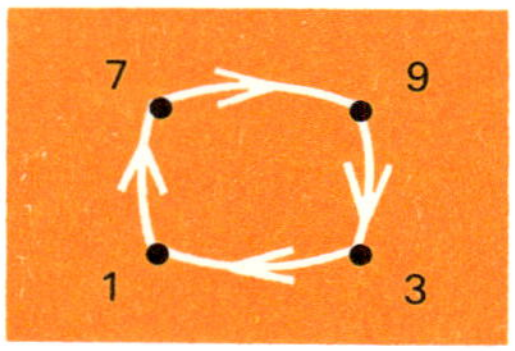

The class might now investigate the end digits of the powers of other numbers less than 10. The end digits form cycles and these can be displayed by arrow graphs. For example, the end digits of powers of 7 are 7, 9, 3, 1, 7, 9, 3, 1, 7, . . . and the graph of this cycle is as shown.

What observations can be made about these cycles?

Investigate what happens to the end digits when powers of a number are written in a base other than 10.

estimating distance

Estimate the length, width and height of the room in centimetres.

How can the information you have be used to get a 'good' approximation of the length, width and height of the room?

This question may suggest some possible sampling technique.

What kind of results would you expect if you asked 6 people in one part of the room and then averaged their results?

This suggestion may come from the class but will almost certainly prompt discussion about the ability to estimate these lengths. Does it make a difference where you stand? Whether you wear glasses? Whether you are male or female? Short or tall? From this discussion a more useful sample might emerge. The results can be collected from the sample and then averaged. The actual measurements could then be found and compared with the sample.

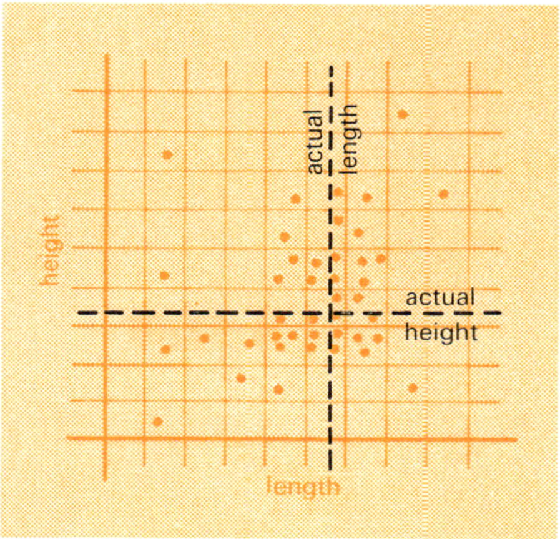

Do those who estimate length/width well also estimate height well?

If all the data was collected at this stage what would be the best way of displaying it in order to answer this question? (No scales are given in the figure as the class interval can only be decided when the range is known.)

More graphs can be drawn which will best suit other questions that might be raised:

Do people estimate horizontal lengths better than vertical lengths?

Does it make any difference to estimates if the observer is outside the room?

Do estimates vary if they are made lying on the floor?

Within what range of accuracy would you be prepared to predict that 50 per cent of estimates would be? 75 per cent?

fractions

Choose any four numbers.

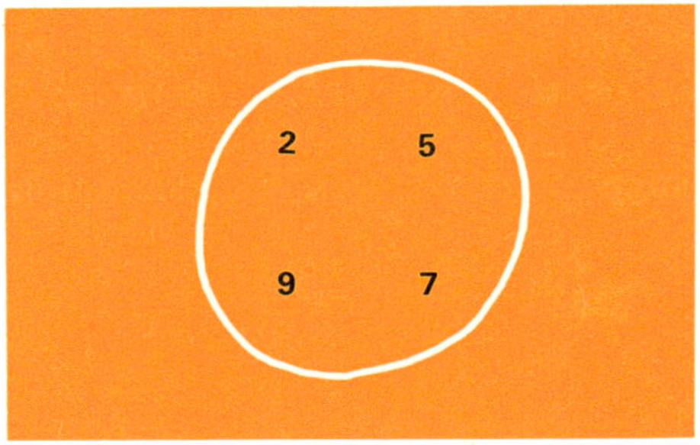

Write a fraction using any two of the numbers.
Write another fraction, another, another, . . .

A question may arise about using the same number twice. If it doesn't. . . .

What fractions can you make using the same number twice?

How many fractions can you make?

How many different fractions can you make?

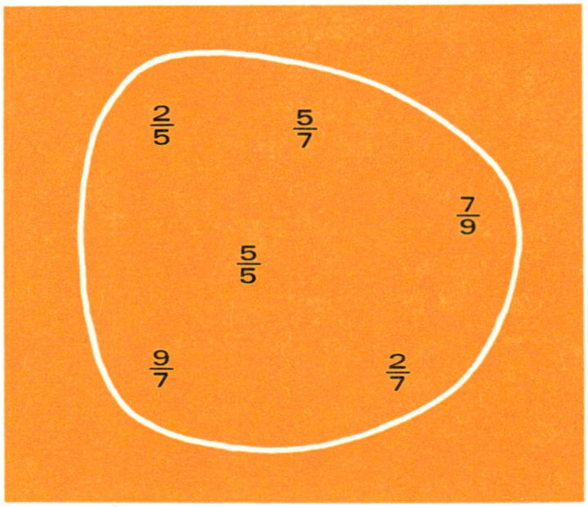

Depending on which four numbers are chosen in the first instance the number of different fractions for any four numbers will be: 13, 11, 9, 7, 5, 3, or 1.

$\frac{2}{2}$	$\frac{5}{2}$	$\frac{7}{2}$	$\frac{9}{2}$
$\frac{2}{5}$	$\frac{5}{5}$	$\frac{7}{5}$	$\frac{9}{5}$
$\frac{2}{7}$	$\frac{5}{7}$	$\frac{7}{7}$	$\frac{9}{7}$
$\frac{2}{9}$	$\frac{5}{9}$	$\frac{7}{9}$	$\frac{9}{9}$

Arrange all the fractions in some form of table.

Several different forms of tabulating are likely to emerge and each will highlight some property of the set of fractions.

Which is the smallest fraction? The largest?
Which fractions are less than 1? Greater than 1? Equal to 1?
Which fractions are equivalent?
Write the fractions in order of size.

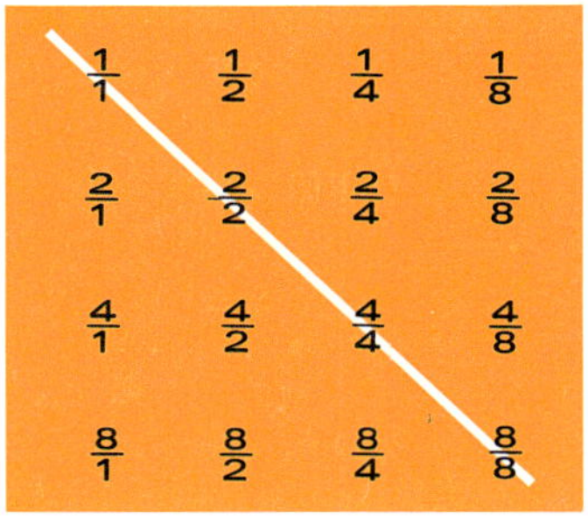

Investigate numbers of fractions for other sets of four numbers.

Consider the four numbers 1, 2, 4, 8. How many different fractions? Equivalent fractions? Fractions greater than 1? Less than 1? Equal to 1? . . . . Which pairs of fractions have a product equal to 1? How many pairs of fractions are there like this? Is it possible to decide on the number of different fractions for any 'n' numbers?

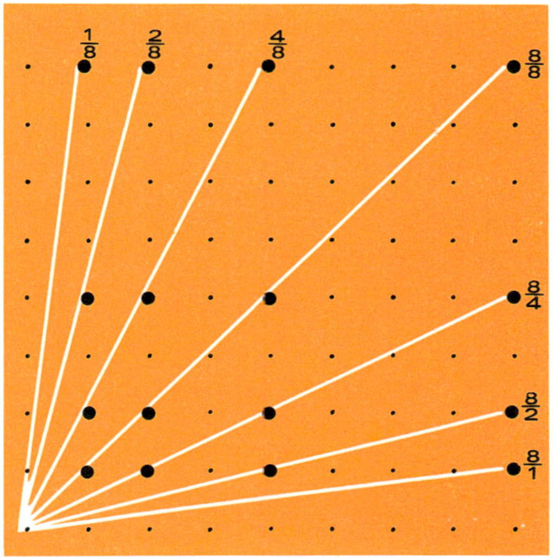

As the fractions have been derived from taking number pairs they can be plotted on a square grid or lattice. (See p. 206). In this way the equivalence is seen as sets of points on straight lines.

The activity starts with four numbers and fractions are made out of these. We might generate other numbers with other operations, $a \star b$ for various $\star$ (e.g. subtraction).

function game

The first part of this lesson is conducted in silence. The blackboard is divided by a vertical line into two sections and it is explained to the class that the left-hand section is for numbers and the right-hand section is for any comments. The comments will be concerned with what is happening in the left-hand section and may be written by anyone as the lesson proceeds. It is also explained that the numbers will be written using two colours of chalk, say white and coloured. Anyone may volunteer to write a white or coloured number at any stage.

From now on—silence.

The teacher holds up the white chalk and a volunteer takes it and writes a white number on the blackboard.

3	4

The teacher, having a specific function in mind, writes a corresponding number in colour.

7	12

The white chalk is offered again and a number written on the board. Again the teacher writes a corresponding coloured number.

Someone else writes a white number but this time the coloured chalk is offered to the class and a volunteer writes a coloured number. (It may be necessary for the teacher to erase an incorrect coloured number.)

4	

More white and coloured numbers are written on the board.

After a while pupils may need to be reminded (by pointing) of the space on the right and encouraged to write something. Comments usually express the relation between the white and coloured numbers but other observations may also appear.

—(white number $-$ 1) $\times$ 2 $=$ coloured number.
—2 $\times$ white number $-$ 2 $=$ coloured number.
—coloured numbers are always even.

Plenty of time is allowed for the writing of comments before the numbers are removed in readiness for another game. The comments are left.

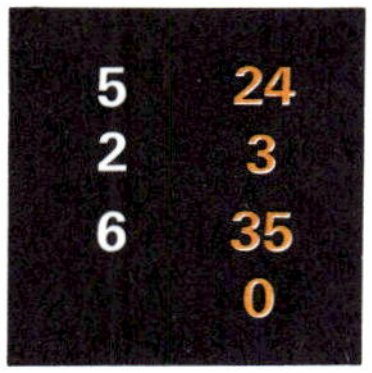

Another game commences with the teacher having a different function in mind. After a few pairs of numbers have been written the teacher might write a coloured number and invite someone to give the corresponding white number.

The lesson might well continue in silence until a third game has been played. After the three games there will be a number of comments in the right-hand section. Attention is now drawn to these.

Can they be written in a simpler way?
Are any brackets necessary?

Where the same function has been written in different ways are the two ways equivalent?
Where some property of the coloured numbers has been observed, for example, 'the coloured numbers are always even', can this be proved?

For another function game see p. 116.

half-turns

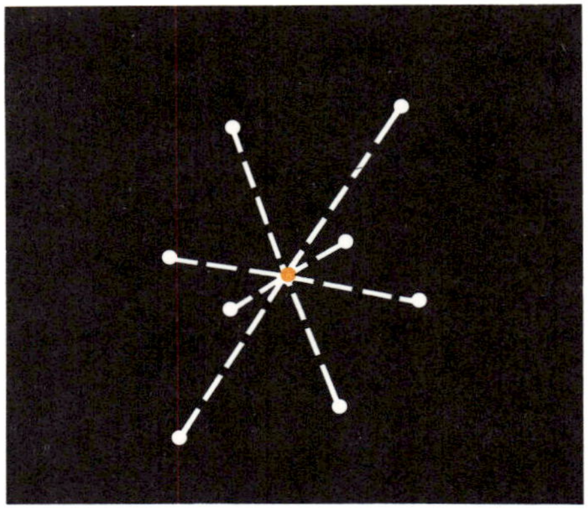

A point somewhere in the middle of the blackboard is marked with coloured chalk. Someone is invited to mark some other point with the white chalk. When this is done the teacher takes the white chalk and marks in the image of the pupil's point in the coloured point.

Other points are asked for and each time the teacher marks in the corresponding image. At some stage the teacher asks someone to put his point in for him. If this is obviously in the wrong place it is rubbed out without comment and the request repeated.

The teacher's rule may or may not need to be verbalized.

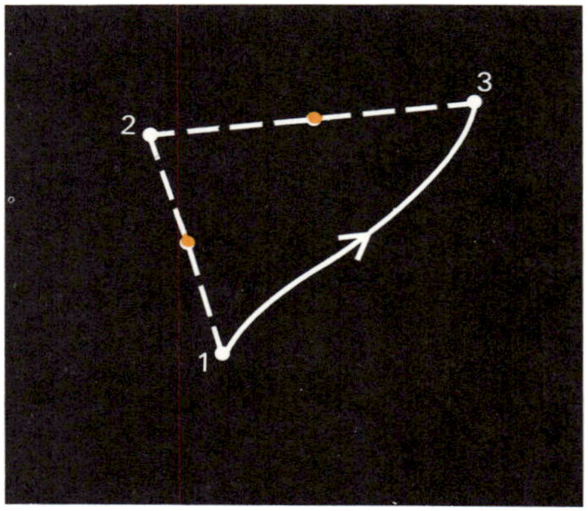

The white points are now erased and a second coloured point is marked. Someone is invited to mark some other point with the white chalk. When this is done the teacher takes the white chalk and marks in the image—in the second coloured point—of the image in the first.

Other points are asked for and each time the teacher marks in the corresponding image of the image (in the same order of coloured points).

The rule may or may not be verbalized.

No particular attention is drawn by the teacher to the order in which he is taking the images.

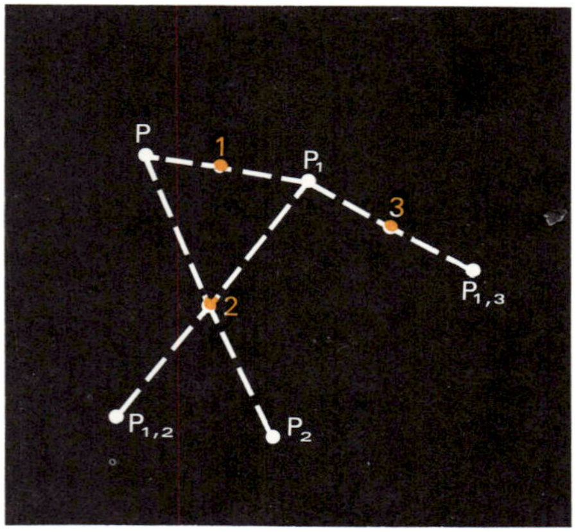

The whole board is erased and three new coloured points are marked. The teacher marks a white point. He invites the next move.

At this stage there may be some discussion. Some way of describing the process will evolve and the order in which the three coloured points are treated may become an issue. All possible orders may be considered and ways of labelling the corresponding (different?) images discussed.

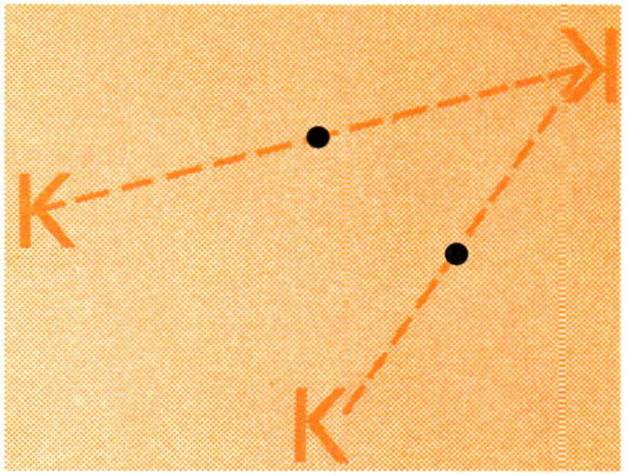

The class may now be invited to work individually or in groups using large sheets of paper and any number of coloured points they prefer. There need be no restriction to small numbers of coloured points though each of them must be used once.

Another direction may be taken by suggesting that a letter be treated as the original element and that it is transformed point by point. Its images in successive numbers of coloured points can be drawn in.

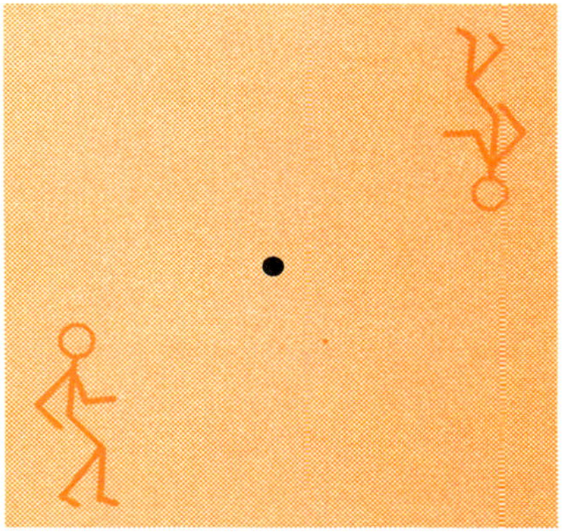

Children readily invent more interesting elements and delight in seeing how these reappear on the page.

How do the elements change? In shape? Size? Position?

Some further issues:

1. What distinctions can be made between images through an odd number and an even number of coloured points?
2. What happens when the coloured points lie in special ways, for example on a straight line or at the vertices of a square?
3. The situation may be reversed. Given an element and one of the same shape and size in a different position can one insert coloured points that will relate the one to the other?
4. The coloured points may be used more than once. Thus with one coloured point we may generate an image A but the image of this image, A^2, is the original point. In this 'algebra' $A^3 = A$ and $A^2 = I$, the identity element. With two coloured points we may generate A, B, AB, BA, ABA and so on. What relations are available in this system? Is there still only a limited number of images?
5. Successive images are taken in coloured *lines*. . . .

halving a line

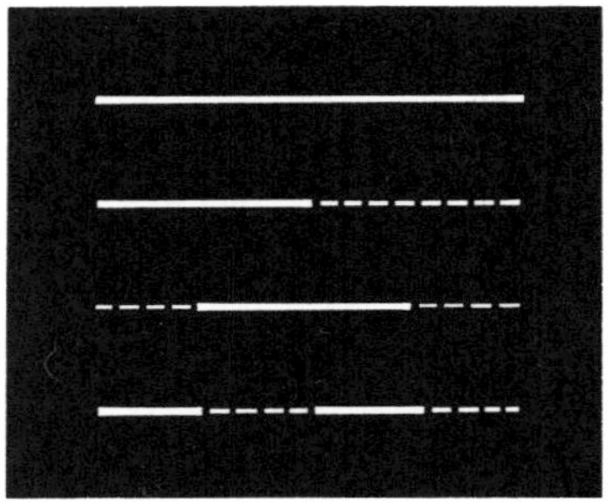

A straight line is drawn on the blackboard. Someone is invited to rub out half of it.

Would anyone have done it differently?

The line is redrawn and other ways of halving it are explored. Someone rubs out various bits of the line that make up half of it. Another insists that the remaining piece must be 'all together'. Another halves the chalky thickness of the line.

The line is redrawn and the teacher proposes that halving be either by rubbing out the left or the right half.

Will someone halve what remains? And now halve that.

A pause. There is now a segment on the board that represents one-eighth of the original line. This may be suggested but not necessarily unanimously agreed. The original line may have to be redrawn.

Can anyone remember which halves we took?

Answers may refer to the bits rubbed off or the bits retained but usually take the form of a sequence such as 'left, right, right'. Agreement may be sought on whether such sequences should refer to bits erased or bits retained. Here we refer to bits retained.

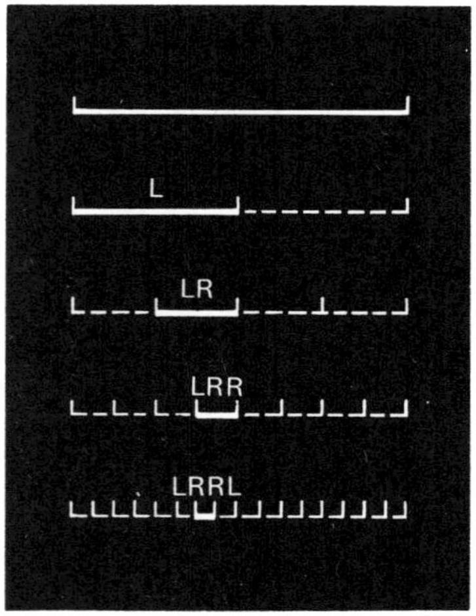

The line is redrawn and a sequence of four instructions requested. This may be recorded on the blackboard, e.g. LRRL where in this case the instructions are read from left to right.

What size is the line that remains?

Whereabouts is it on the original line?

Some sort this out mentally; others make marks on paper. There may be various ways of describing the second answer.

—It is the bit just before the one before the middle.

—It is the seventh along, no it's the sixth.

It is interesting to attack the problem with eyes closed.

Imagine the line. . . . Keep the right; now the left; now the right; hold it. Open your eyes. . . .

The problem can be reversed. The line is redrawn and divided roughly into eighths. A particular segment is chosen and the class is invited to give the sequence of instructions that would leave that segment.

What about the other segments?

What about sixteenths?

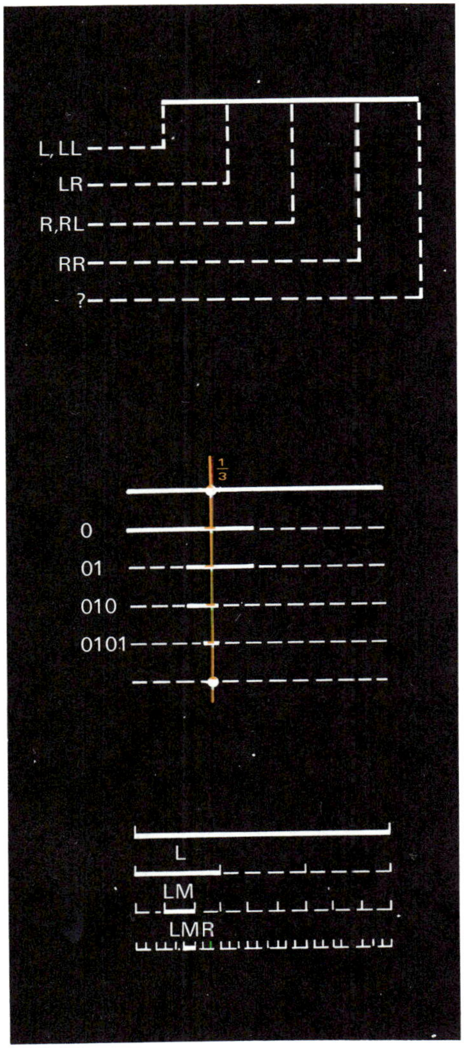

Some further issues:

1. Can we go on halving?

2. The left-hand point of any segment arising from a succession of halvings is identified by the sequence of instructions. Thus any sequence of L's and R's will refer to a point of the line. Can a point have two such labels?
   Are there any points that cannot be labelled?

3. Suppose 0 and 1 are used to label 'Left' and 'Right' respectively. What sequence of 0's and 1's labels the left-hand point of the original line? The right-hand point? The mid point? A point of trisection?

4. Express 1/3 as a binary fraction.

5. Consider trisecting a line by retaining either the left-third, the middle-third, or the right-third.
   What does LMR retain?
   What sequence of L's, M's and R's labels the mid point of the line?

intersecting circles

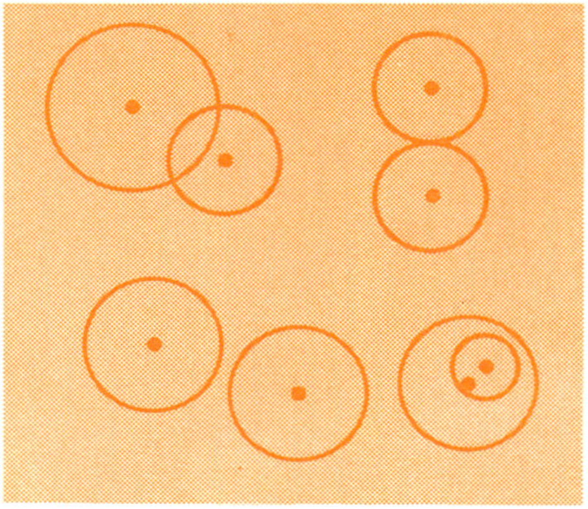

Draw two circles and mark their centres.

Examine and discuss the various situations which have been drawn.

The relative positions of the centres, the number of points of intersection, 'touching' circles and the area of overlap are some of the observations usually discussed.

Now draw a circle.

Draw another circle the same size which intersects the first circle.

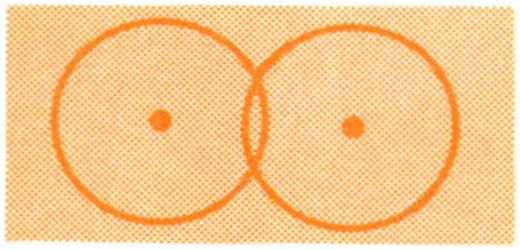

How did you know the circles would intersect?
If the centres were further apart would they intersect?
How much further apart?

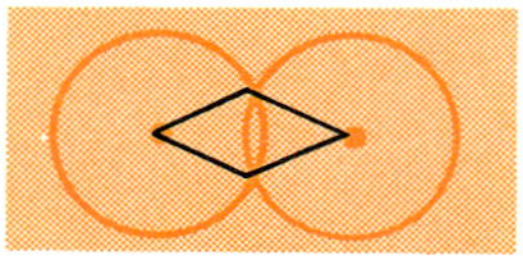

Now join the points of intersection to the centres.
What can we say about the shape formed by these lines?

Draw another figure with the centres nearer to each other.

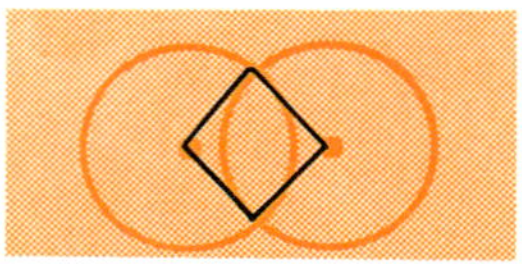

What has happened to the rhombus?
Will we always get a rhombus?

Draw more cases.

When does the rhombus become a square?

Draw the case where the circumference of one circle passes through the centre of the other.

What can we say about this rhombus?

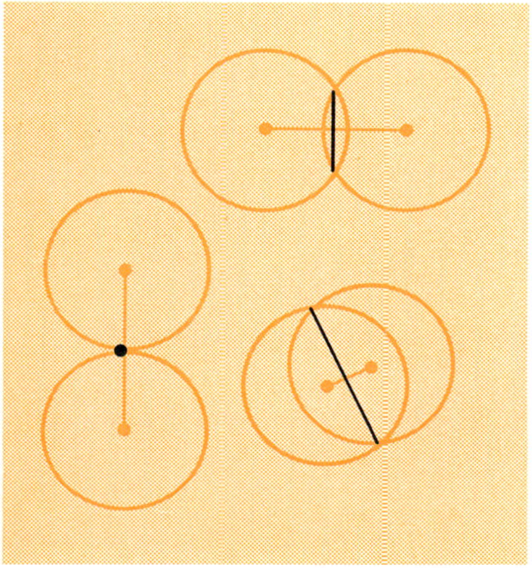

On the figures you have drawn join the two centres, and join the two points of intersection.
What do you notice about these two lines?

—They bisect each other.
—They cross at right angles.
—When they are the same length it is a square.
—They bisect the angles.

The relation between their lengths might be investigated.

Now pupils can try constructing a rhombus to satisfy certain conditions: e.g. given the length of side and one diagonal; given the lengths of the two diagonals.

Some standard compass constructions readily follow:
the bisection of a straight line.
the bisection of an angle.
the perpendicular bisector of a straight line.

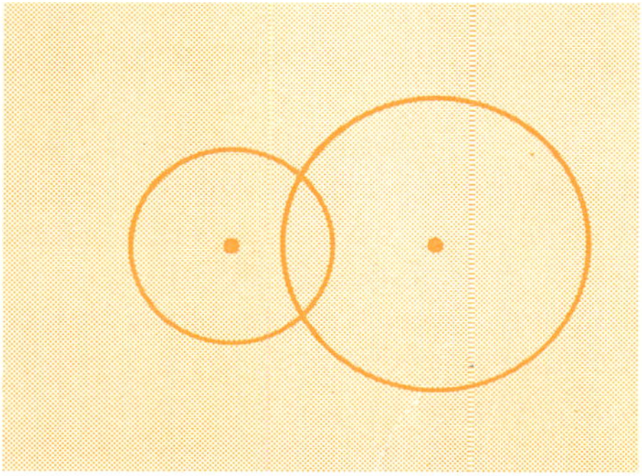

In the same way investigate two intersecting circles which are not the same size.

This theme makes a suitable subject for a filmloop (for information on filmloops, see p. 191).

logarithms

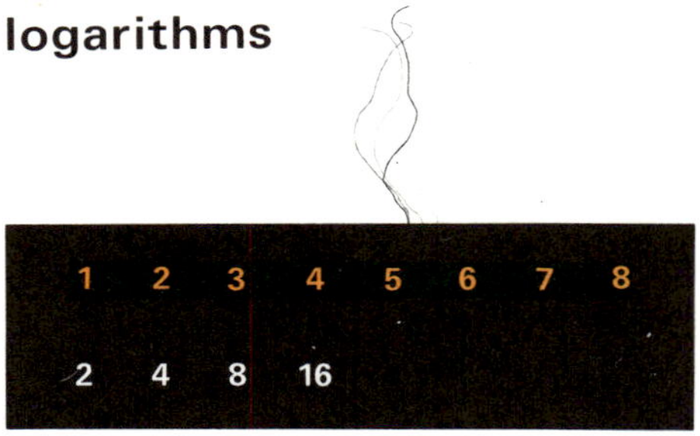

Coloured and white numbers are written on the blackboard as shown.

What white number should be written under the coloured 5?

Under the coloured 6?

What white number should be written under coloured 8?

If I give you a coloured number can you give me the white number?

If I give you a white number can you give me the coloured number?

What is the coloured number for a white 3?

What is the coloured number for a white 6?

At this stage it is likely that the coloured numbers suggested are $1 \cdot 5$ and $2 \cdot 5$ respectively. For the time being these are accepted without comment and written on the board.

The teacher using white numbers then performs the calculation $4 \times 8 = 32$ (this is written on the blackboard). Someone is asked to come to the blackboard and perform the same calculation using coloured numbers. It is quite likely that the calculation performed will be $2 \times 3 = 6$; in which case it will be necessary to comment: Why did you write 2? 3? But the coloured number for 32 is not 6!

In this way the homomorphism between the white numbers with multiplication, and the coloured numbers with addition is carefully established. The teacher using white numbers performs the calculation $4 \times 16 = 64$ and a pupil, using coloured numbers will perform $2 + 4 = 6$.

The teacher now writes in white $3 \times 6 = 18$ and invites someone to do the corresponding coloured calculation. Using the values suggested for 3 and 6 this will be $1 \cdot 5 + 2 \cdot 5 = 4$.

Pupils will quickly observe the discrepancy with 4 being the coloured number for white 16. This suggests that the guesses for 3 and 6 are not satisfactory.

If we consider $6 \times 6 = 36$ and the corresponding $2{\cdot}5 + 2{\cdot}5 = 5$ we see that the value 2·5 suggested for 6 is too high. In the same way we consider $3 \times 3 = 9$ and $1{\cdot}5 + 1{\cdot}5 = 3$.

The task now might be to find consistent coloured numbers for white numbers from 1 to 10.

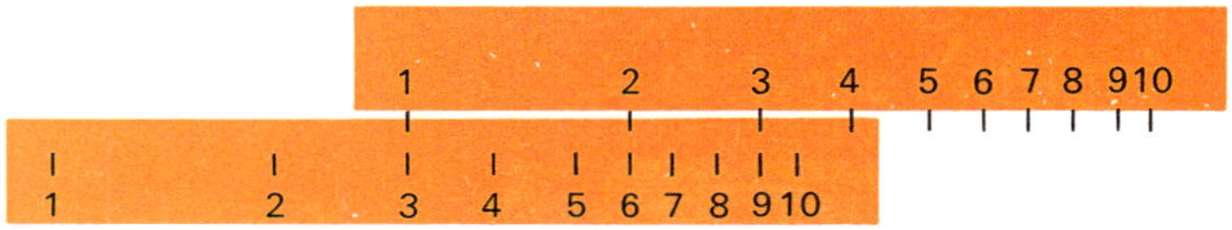

The coloured numbers are, of course, the logarithms of the white numbers to the base of 2.

The numbers may be placed on two strips of card, with corresponding scales, to form an elementary slide rule.

Another investigation starts with white numbers 10, 100, 1000 and so on, and red numbers, say, which are the corresponding powers of 10. The red numbers will be the logarithms of the white numbers to base 10. One task is now to find consistent red numbers for the white numbers from 1 to 20 without using tables or graphs. This can be done with reasonable accuracy though some ingenuity may be required to complete the list. In the first instance we may note that 2^{10} is 1024 which is approximately 10^3, so that log 2 is approximately 0·3.

For a number such as 17, it may be noted that $17^3 = 4913$, so that log 17 can be calculated from the value chosen for log 2 by expressing 17^3 as approximately $10^4 \div 2$, or from the value chosen for log 7 by expressing 17^3 as $10^2 \times 7^2$.

group structure

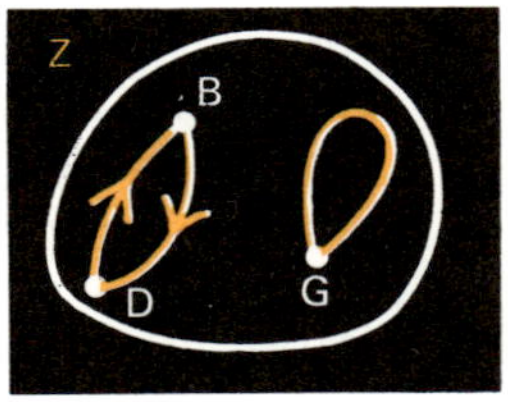

Three people, Bill, Dorothy and George stand in a row facing the class. The teacher draws an arrowgraph on the board.

After the class has had an opportunity to discuss the arrowgraph the three people at the front change order as in the graph. The class is invited to give a symbol for this instruction (say Z). The three are restored to the original order.

Someone is invited to draw another arrowgraph, and draws, say, the one shown. The three are asked to do what this graph says, and again a symbol is given for this instruction (say Y). The original order is again restored.

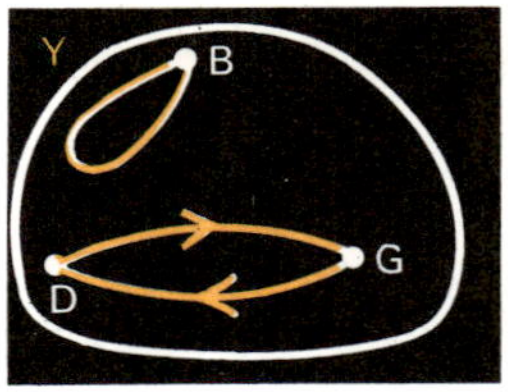

What happens if we do a Z and then follow this by a Y?

The three do a Z and follow this by a Y and the new order is observed.

Suppose we wish to get Bill, Dorothy and George in this order with just one instruction, could someone draw the arrowgraph for it?

This instruction is given the symbol X and it is also recorded that Z followed by Y is X (ZfY = X).

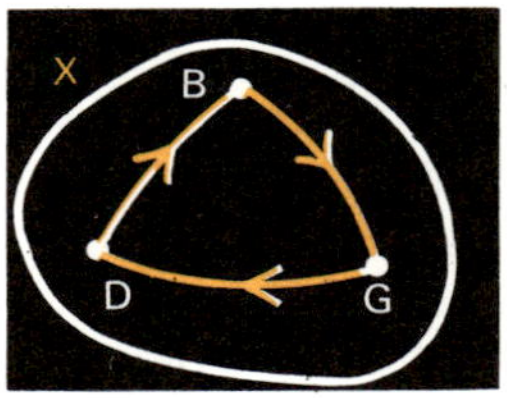

Now what could we try?

—ZfX; YfX; YfY; XfZ and so on.

Let's try ZfX. What will happen to Bill? to Dorothy? to George?

The instructions are performed, and again the order is observed.

—This is the same as doing Y.

So it is not necessary to have a new arrowgraph for ZfX, but the result is recorded.

ZfX = Y

The class now use pieces of card or counters to find the results of combining other instructions.

If the result of a combination produces a new order then a new instruction is established. This, in turn, provides further combinations to be tried. If the result of a combination produces an order previously obtained then no new instruction is necessary, but the result is recorded. Such combinations as ZfZ show the need for a 'do nothing' instruction.

Finally the results of combining instructions are collected together in a two-entry table and the structure of the table observed.

mark and rub

A girl, say, has the chalk and is asked to make a mark on the board every time she hears the instruction 'mark'. A boy has the board rubber and is asked to wipe off the last single mark made on the board every time he hears the instruction 'rub'.

Mark, mark, rub, mark, rub, mark.

—you are left with two marks.

—you could get two marks in a different way—mark, mark, rub, mark. . . .

Could we get just one mark left?

—yes, mark, rub, mark.

How can we write down what we are doing?

Can you give me some instructions?

—MRMR . . . there is nothing left.

Can we leave the board empty with other instructions?

—MMRR . . . RMRR . . . RMMR—no you fool, that will leave a mark.

—MRRR . . . RRRR . . .

—any number of rubs together with no marks leaves the board empty.

—you can have as many R's in front as you like. . . so we could leave them out . . . no it wouldn't be quite the same . . . it makes a difference if you put them at the end.

What about the M's?

—they make a difference.

Can we write down instructions which will leave one mark on the board?

—MRMRMRM . . . MMMRR . . . and so on.

—Couldn't you write 4R for RRRR—it would be quicker.

—Why not R4?

The teacher may use various symbolisms at this point or can encourage the use of R^4.

Expressions such as $R^6MRMR^7M^7RMR^8$ may now be written on the blackboard.

—Nothing.

—I've got one . . . $M^{20}R^{10}MR^6$.

—That's five marks.

An 'algebra' is developed.

monodivisors

Someone in the class is asked to give a number (he gives, say, 9). Someone else is asked for a number (he gives, say, 7).

The two numbers are written as a pair on the blackboard.

The class is invited to make comments about this pair.

—Added together they make 16.

—The difference between them is 2.

—If you multiply them you get 63.

—They are both odd.

The teacher now writes another pair of numbers on the board, and comments are invited. Apart from suggestions like those previously mentioned it is likely that someone will observe that 4 will divide exactly into each number.

Are there any other numbers that will divide exactly into each of the pair? Do we count 1?

Attention is now turned to pairs of numbers having only one common divisor—namely 1. Pairs are restricted to the second number being less than the first, as, for example, (12, 5) and (5, 12) are both pairs of the type required so only (12, 5) is counted.

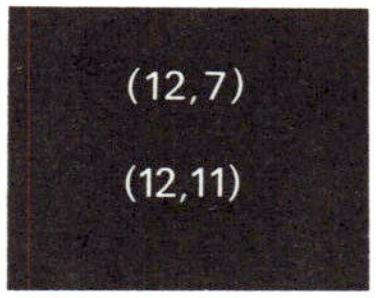

If we write 12 as the first number, what numbers less than 12 could be written for the second number so that the pair have only one common divisor?

The pairs are collected on the board. It is helpful to establish a name for such pairs of numbers. One class called them *monodivisors* and this name is used here.

How many monodivisors has 12?

How many monodivisors has 9?

The numbers 1 to 100 are shared among the class and a table of 'numbers of monodivisors' is formed as a class activity.

Various patterns are observed.

—Some look at the powers of 2 in the monodivisor table and detect a pattern.

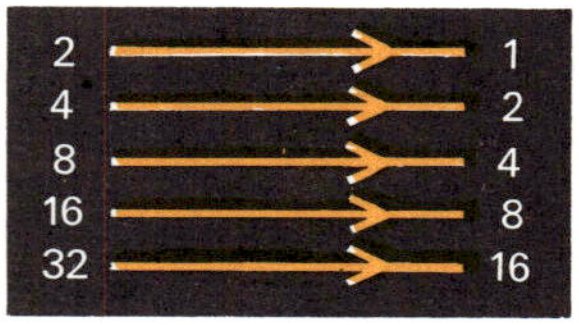

—Others take multiples of a number and look at the corresponding monodivisor pattern.

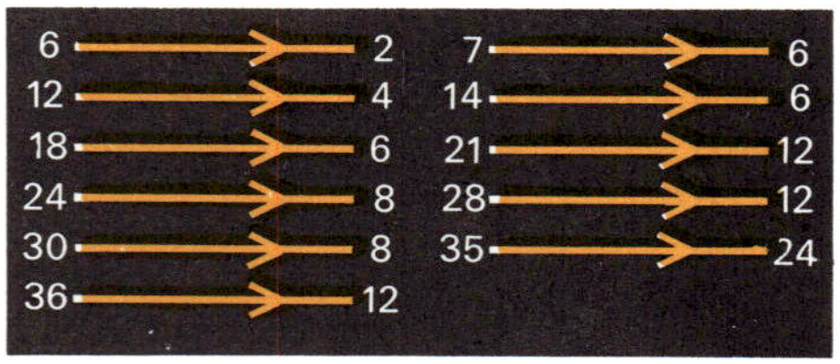

(72,1)
5) 4
7) 2
11) 4
13) 2
17) 4
19) 2
23) 4

0	1	2	3	4	5	6
1	2	3		5		7
		4		8		9
		6		10		14
				12		28

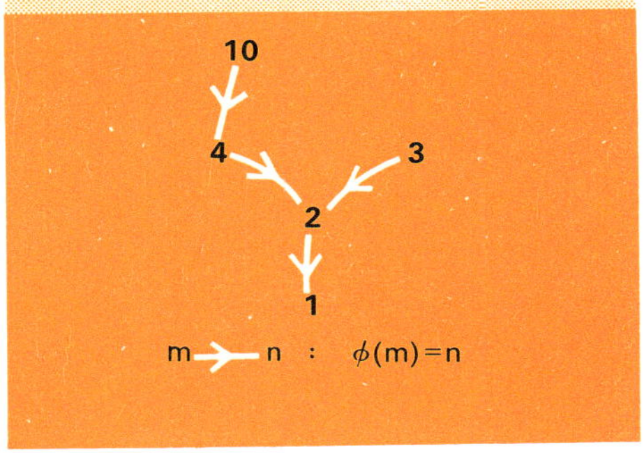

—Some go back to look at the monodivisor pairs for, say, 72, and observe patterns in these.

Numbers having the same number of monodivisors might be collected together in a table. This reveals that no number, except 2, has an odd number of monodivisors.

Why is this?

What else does the table reveal?

Others might try operating with the monodivisor numbers. For example: $35 = 5 \times 7$. 5 has monodivisor number of 4, and 7 has monodivisor number of 6. Multiplying these together we obtain 24 which is the monodivisor number of 35.

For which numbers does this apply?

Some further issues:

1. Place a coloured peg in a piece of pegboard and then insert pegs of another colour in such a way that there are no holes in a straight line between them and the coloured peg. The co-ordinates of these pegs are monodivisor pairs.
   Investigate the pattern formed by these pegs. Cf. p. 219.
2. Consider duodivisors, i.e. pairs of numbers (e.g. 12, 3) that have two and only two common divisors (namely 1 and 3).
3. The number of monodivisors of n is usually denoted by $\varphi(n)$ and called Euler's function after the mathematician who first investigated them. For a prime number p, $\varphi(p) = p - 1$. For example, $\varphi(13) = 12$ because all 12 numbers less than 13 have only one common divisor with 13. Establish other formulae using this notation.
4. Construct an arrowgraph where an arrow from m to n means that $\varphi(m) = n$.
5. The $\varphi(n)$ monodivisors form a group under multiplication, mod n.

movements

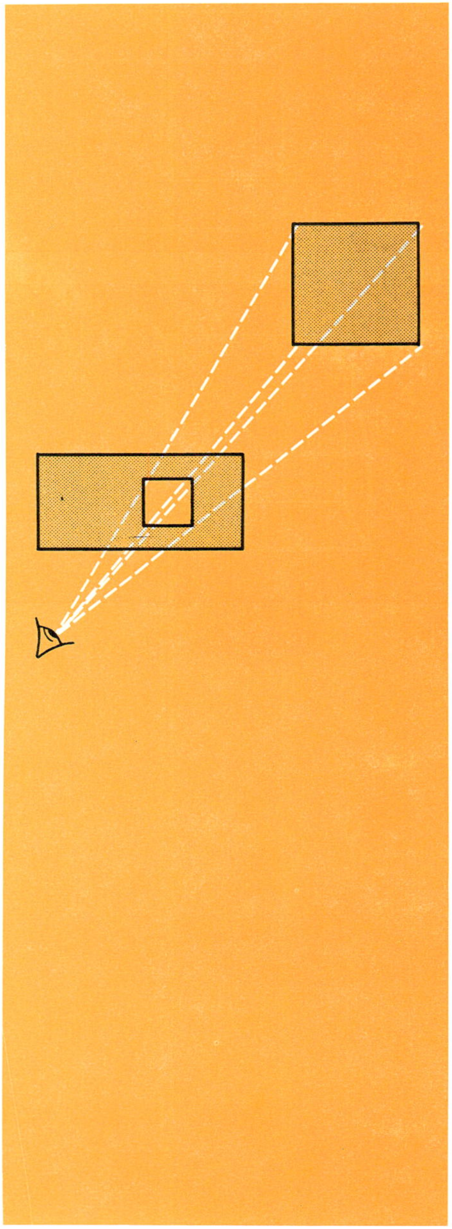

Each pupil is given a large sheet of paper with a square hole of about 2 cm side cut out from the middle (or they are asked to cut out the square themselves). The teacher has a piece of card or hardboard about 20 cm square with a small block of wood nailed or stuck on one side. He uses the block as a handle holding up the square so that the plain side faces the class.

He asks the class to raise their papers and frame his square exactly into their hole.

(*Some people are closer to their papers than others, which ones?*)

Can you keep the square framed in your hole?

The square is moved in various ways. This motion may be restricted in some way.

—the square may turn about its centre.
—it recedes or moves forward.
—it slips sideways.
—it turns about a line through its centre.
—it combines some of these motions.

Can you tell what the square is doing?
Is your hole doing something similar?
What shapes can you see?
Will someone come and move the big square?
Can we do this with other shapes?

Such questions indicate some of the things the teacher may have in mind but may be left implicit. The situation is interesting and can be amusing. Pupils like breaking up into pairs and experimenting on their own. Useful informal experience is gained that will be available for later more formal discussions of translations, rotations, symmetries and so on.

number transformations

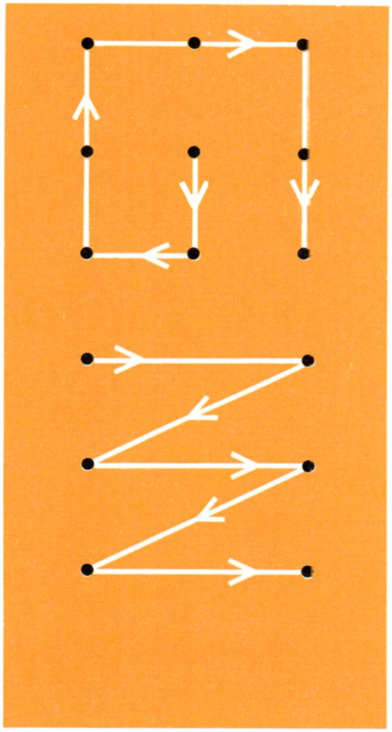

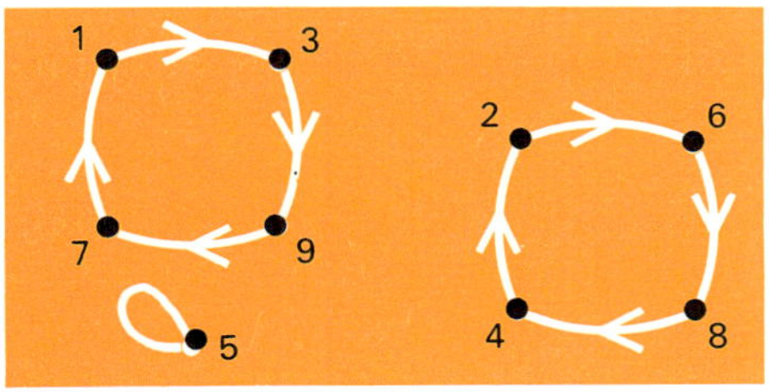

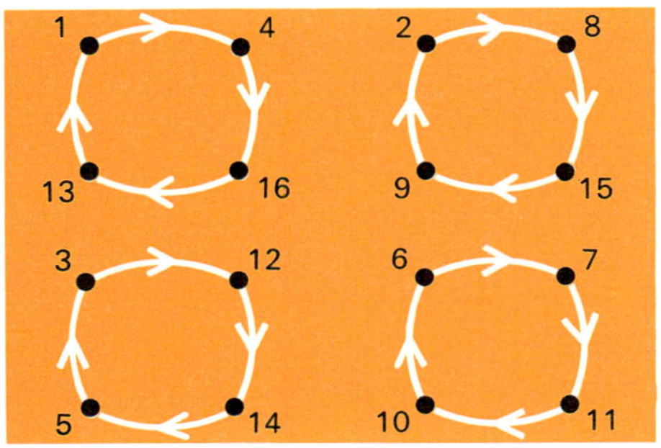

A 9 pin geo-board is held in front of the class. Pupils are invited to suggest ways in which the pins could be numbered.

Suggestions usually include 'lines', 'spirals', 'diagonals', and so on as indicated.

One particular method of numbering is chosen, say, 'lines from left to right'. Various questions are asked to establish that everyone is mentally numbering correctly. (Placing numbers on the geo-boards detracts from the value of what follows).

Which pin is 6?

Which pin is 4?

Which pin is this?

The board is now given a quarter turn in a clockwise direction.

Where has 1 gone?

Where has 2 gone?

Where has 7 gone?

The mapping may be shown by arrows.

1 has gone to 3; what has happened to 3?

So what has happened to 9?

So what has happened to 7?

In this way cycles of numbers are produced.

Draw graphs of all the cycles.

What do you notice?

—One cycle contains odd numbers and the other even numbers.

—The numbers in the cycles add up to 20.

—The numbers diagonally add up to 10.

It is possible that some pupils will connect the numbers in the cycles with multiplication by 3 (mod 10).

The investigation might now be extended to a 16-pin geo-board, numbering in the same manner.

The graphs are as shown. The total of the numbers in each cycle is 34 and the numbers in a cycle are obtained by multiplication by 4 (mod 17).

What happens if the boards are rotated anti-clockwise?

What happens if boards are numbered in the other ways which have been suggested?

next to . . .

A coloured peg is inserted into a piece of pegboard.
Place another peg next to the coloured one.
Various possibilities are available:

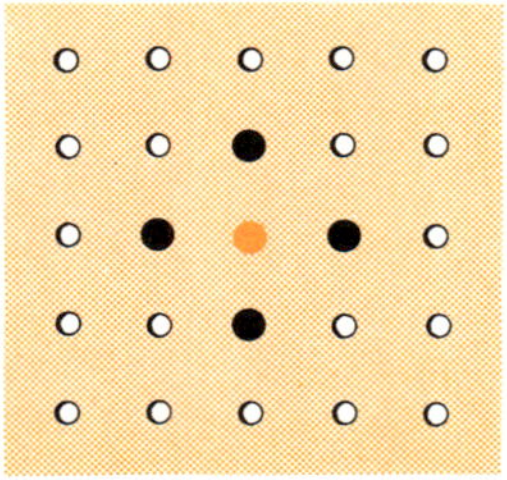

(a) For one pupil the phrase 'next to' may mean 'immediately adjacent in a rook's move' though he may express this in other ways. All may agree with him in which case the teacher may proceed using this rule or invite other possibilities.

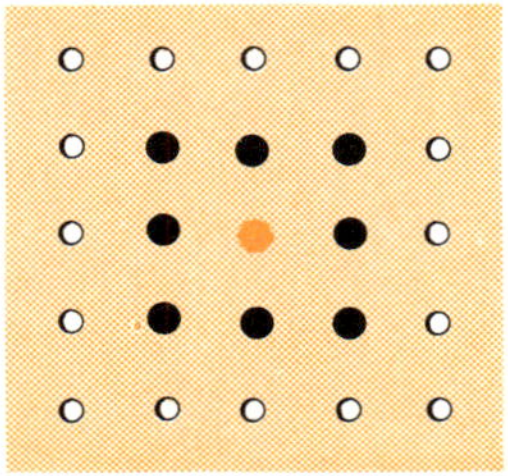

(b) Another pupil may interpret in a similar way but using a king's move.

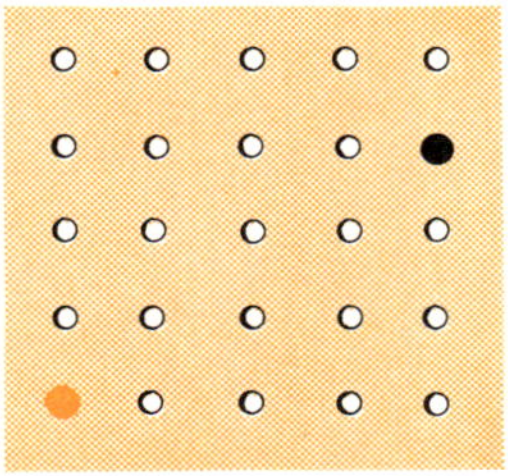

(c) Another places the second peg well away to one side of the board.
—there are no holes in a straight line between my peg and the coloured peg.
Could this be a possible meaning of 'next to'?

Discussion of such possibilities makes an excellent starting point precisely because there are many directions to take. It is of course also a matter of exploring these directions further. What are the implications of any agreed interpretation? In each case further questions may indicate a particular inquiry.

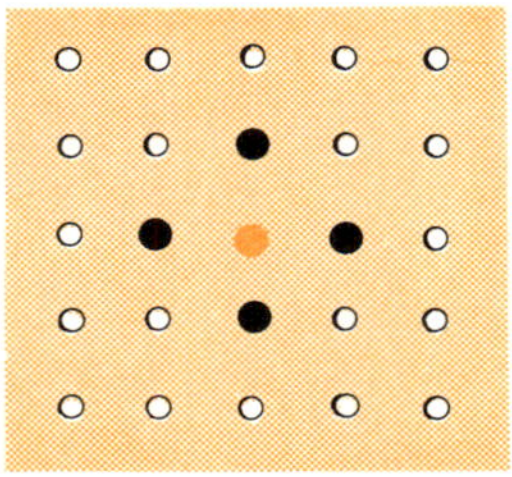

1. Interpret 'next to' to mean 'immediately adjacent in a rook's move' (the reader not familiar with chess moves should consult the accompanying figure).

0	1
1	4
2	8
3	.

How many pegs are there next to the coloured one?
(These are labelled 1 in the figure).
How many new ones next to one of these?
And next to one of these?
What sequence is generated?

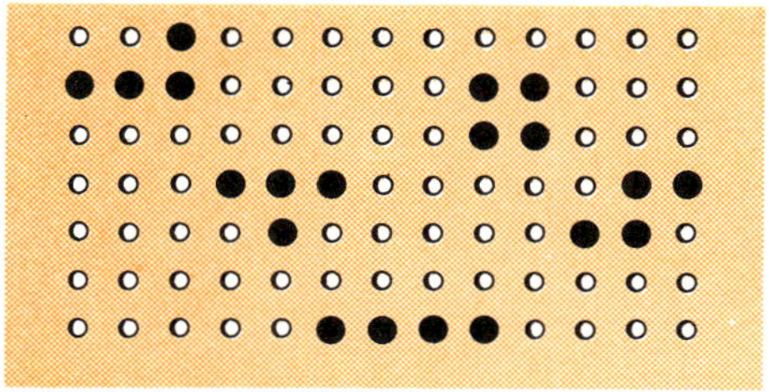

With the same rule place a third peg next to one of two pegs that are next to each other. And a fourth one next to one of the three.

How many different arrangements of pegs built up in this way can arise.

Investigate further. (See p. 118, p. 231).

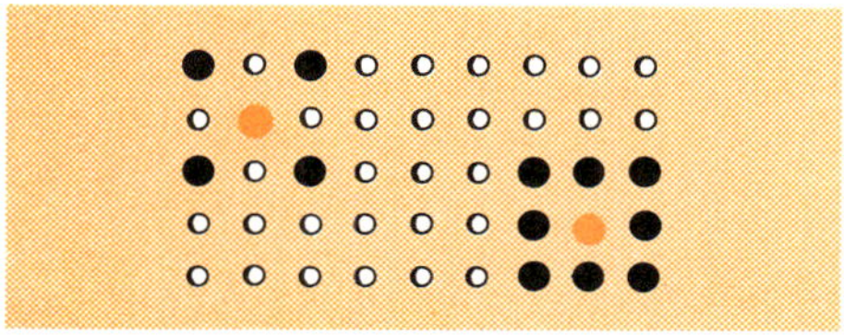

2. The preceding questions may be raised with the bishop's move interpretation. . . .

3. Or with a combination of the two rules, i.e. using the king's move interpretation.

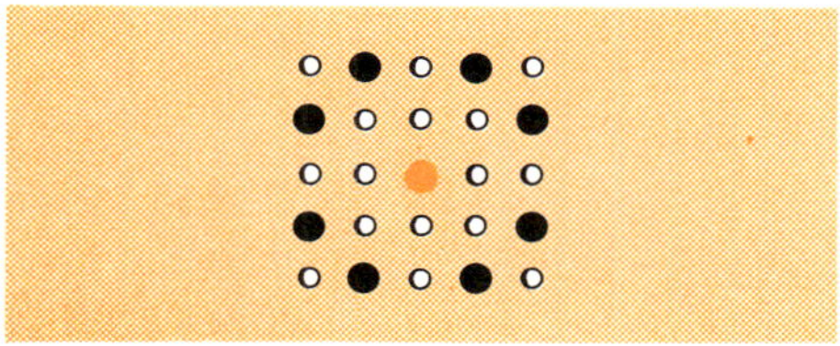

4. Interpret 'next to' to mean 'immediately adjacent in a knight's move'.

The holes of the pegboard—or squares of squared paper—can be numbered with the least number of knights' moves required to reach the hole from the coloured peg.

What patterns arise?

Consider arrangements of four pegs as in item 1 but using the knight's move rule.

On squared paper the corresponding patterns are disconnected.

Investigate further.

5. Interpret 'next to' to mean 'any position that can be reached in a queen's move'.

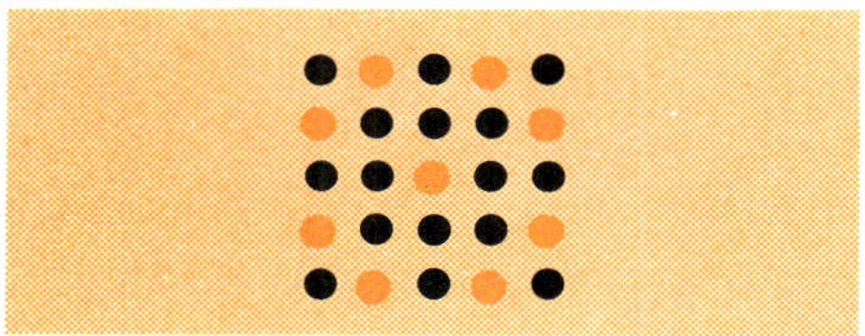

Insert another coloured peg as close as possible to the first one without being 'next to' it. Place further coloured pegs in all similar positions.

nine times table

1 × 9 = 9	
2 × 9 = 18	
3 × 9 = 27	
4 × 9 = 36	
5 × 9 = 45	
6 × 9 = 54	
7 × 9 = 63	
8 × 9 = 72	
9 × 9 = 81	
10 × 9 = 90	
11 × 9 = 99	→ 18
12 × 9 = 108	
13 × 9 = 117	
14 × 9 = 126	
15 × 9 = 135	
16 × 9 = 144	
17 × 9 = 153	
18 × 9 = 162	
19 × 9 = 171	
20 × 9 = 180	
21 × 9 = 189	→ 18
22 × 9 = 198	→ 18
23 × 9 = 207	
24 × 9 = 216	
25 × 9 = 225	
26 × 9 = 234	
27 × 9 = 243	
28 × 9 = 252	
29 × 9 = 261	
30 × 9 = 270	
31 × 9 = 279	→ 18
32 × 9 = 288	→ 18
33 × 9 = 297	→ 18

Write the 9-times table.

What do you notice?

—The 'tens' numbers start at 1 and go up.

—The 'units' start at 9 and go down.

—The 'tens' don't always go up—9 comes twice, 18 comes twice, twenty-seven. . . .

What other numbers will recur in this way?

Can you see any way of remembering a number in the 9-times table?

—The digits of the number add up to 9.

Do they always?

—No, 9×11 doesn't, 9×21 and 9×22 don't; 9×31, 9×32, 9×33, don't. . . . But they do if you add again. 99: $9+9=18$, $1+8=9$. 189: $1+8+9=18$, $1+8=9$.

How often does the digit-sum of 18 occur?
When will it occur again—how many 18's will there be?

Are there any digit-sums other than 9 and 18?
What do you expect them to be?

Can you find a way of deciding on the result of $9 \times n$ quickly?

This may lead to some ideas like: If n falls between 1 and 9 the tens number will be $(n-1)$ and the digit sum 9. If n is between 11 and 20 the tens number will be $(n-2)$; between 21 and 30: $(n-3)$. . . . The digit sum will always be 9 unless it falls into the pattern of 18 (above), e.g. 9×67 will give a tens of $(67-7)$ and a digit-sum of 9, so the product will be 603.

Does this ever break down?

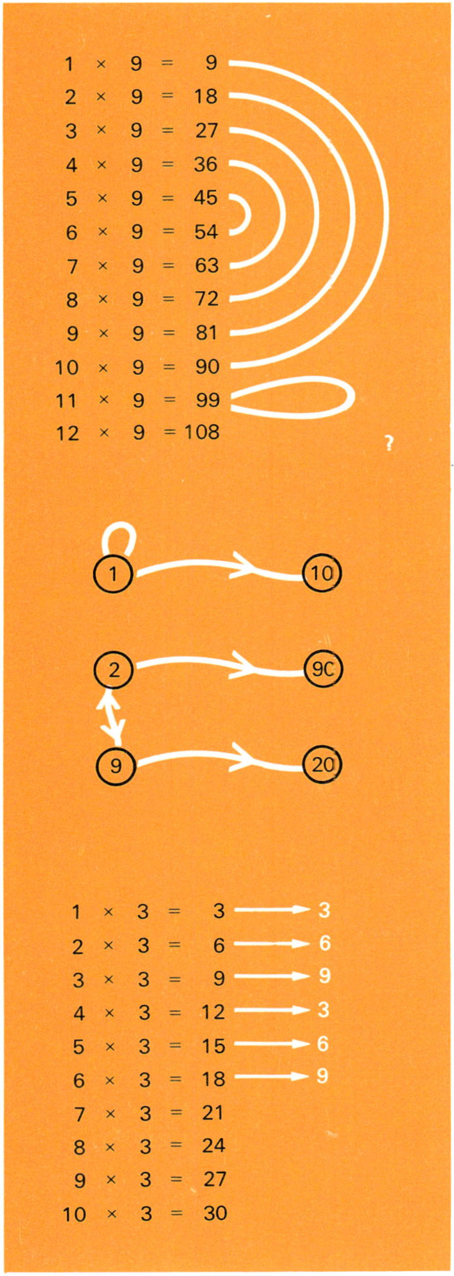

It is likely that with further questioning the reversibility of digits will be noticed. If not:—

What do you notice about the digits of

4×9 and 7×9? (36 and 63)

2×9 and 9×9? (18 and 81)

As 7×9 produces a result which reverses the digits of 4×9 we could say that 7 is the reversal of 4: $R4 = 7$. In the same way $R2 = 9$.

What is R12? $12 \times 9 = 108$. Reversing the digits gives 801.

$? \times 9 = 801$, $89 \times 9 = 801$

so $R12 = 89$.

By continuing the 9-times table try to find some pattern of reversals.

It might be worth studying reversal operators using arrowgraphs.

	$1 \times 9 = 9$	$R1 = 1$
	$10 \times 9 = 90$	$R10 = 1$
or	$2 \times 9 = 18$	$R2 = 9$
	$9 \times 9 = 81$	$R9 = 2$
	$90 \times 9 = 810$	$R90 = 2$
	$20 \times 9 = 180$	$R20 = 9$

Investigate other tables for various patterns.

A class discusses the 3-times table:

—It keeps adding up to 3, 6, 9.

—It goes 3, 6, 9, then 2, 5, 8, then 1, 4, 7, and then stops.

—The tens digit changes every 3.

—There is 12 and 21 with a gap of 3 places, 27 and 72 with a gap of 15. . . .

paper folding

Each pupil needs a strip of paper about 30 cm by 5 cm.

Fold the piece of paper in half.
Now fold it in half again.
Open out the piece of paper.
What do you notice?

This is likely to produce comments on such things as number of folds, parallel lines, number of rectangles, upward and downward folds.

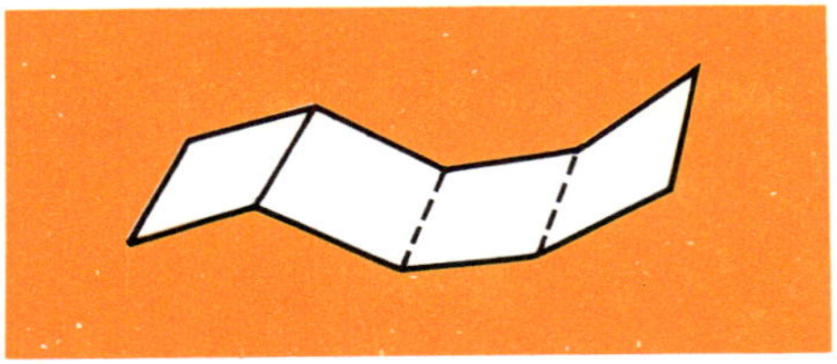

Place your paper like this (opposite).
Reading from left to right what are the folds?
In this case they will be upward, downward, downward.

Can you fold a piece of paper to get a different combination of folds?

Fold your piece of paper again as it was.
Now fold your paper in half again.
Sit on it and don't open it up until you are told!
How many folds do you think this time?
Can you give a reason for this?
How many will be upward?
How many will be downwards?
Now open it out and see if you were right.

number of times folded	number of folds
1	1
2	3
3	7
4	15
.	.
.	.
.	.

The situation might now be investigated in different ways.

1. Investigate the number of folds if folding of the paper is continued. What is the relation between the number of times folded and the number of folds?

number of folds	downward folds	upward folds
1	1	0
2	2	1
3	–	–

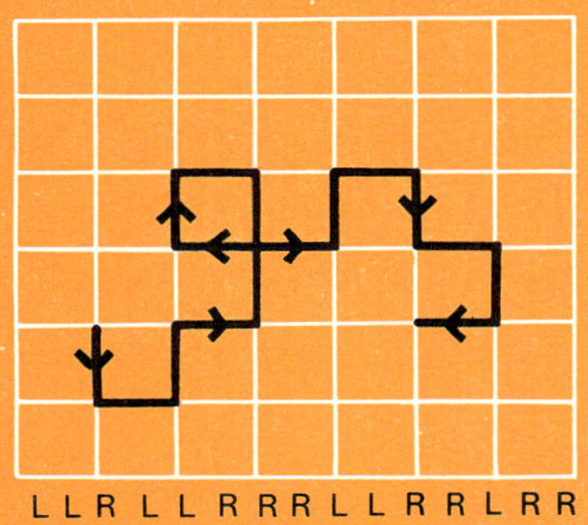

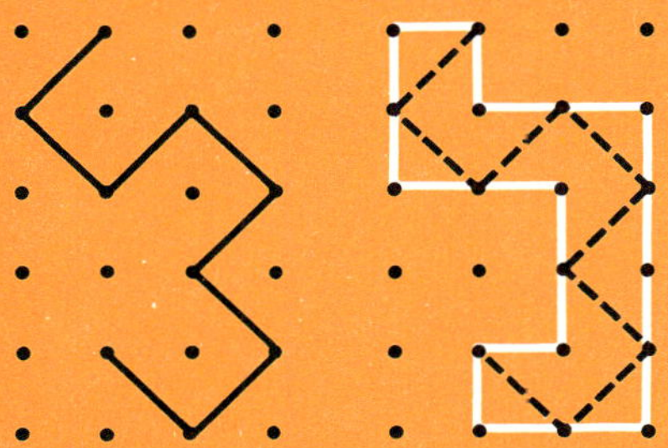

2. A new piece of paper will be required and it must be clearly established which way the folding is to be done, e.g. lay the paper on the desk and always fold left over right. Open out in the reverse way and read your pattern of folds from left to right.
After 1 fold—down.
After 2 folds—up, down, down.
The class can be encouraged to symbolize.
Investigate the patterns obtained if folding is continued.
Can you say what the folds would be after 5 folds? 10 folds? 100 folds? *n* folds?

3. Folding in the same way as above, investigate the total number of folds, the number of downward folds and the number of upward folds.

4. Investigate the different combinations of folds which can be obtained by folding the paper in different ways.

5. After a number of folds the piece of paper can be opened out so that there is a right-angle at each fold. The resulting pattern can be plotted on squared paper and can be coded in terms of the left or right turns taken at each point as a route is traced from one end to another.
A sequence of letters, L or R, produces a 'word' which describes a particular fold. What is the pattern which generates such words?

6. Plot the zig-zag made by the folds with right-angle turns, as in the previous section, on the *diagonals* of squared paper. Each 'step' is now the diagonal of a unit square of the paper.

Shade or colour in such squares. Investigate the resulting shapes—e.g. do they tile?

parallel lines

Draw 2 parallel lines on a piece of paper.

Draw a line on a piece of acetate paper. (Cf. p. 158)

Place the acetate sheet over the paper as shown.

What do you notice?

Move the acetate. What do you notice?

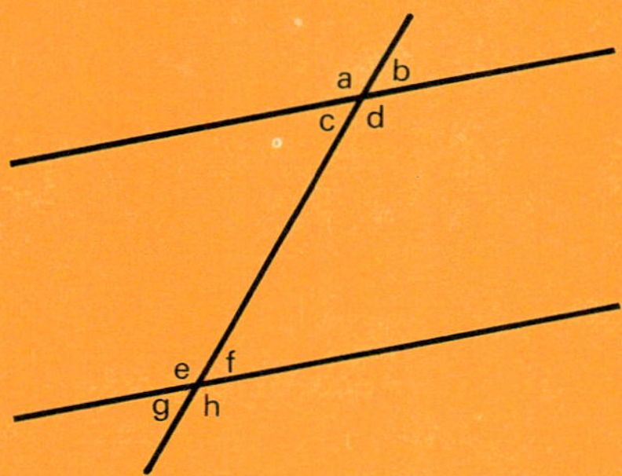

Point to an angle.
Find another angle like it.
Find another, another, . . .
How many angles are there like the one you chose?

Find a different angle.
How many of these?

How many different angles are there?

During this activity a diagram with the angles marked would be useful.

A language could be developed to describe equal angles—either the pupils' or the official language (vertically opposite, corresponding, alternate).

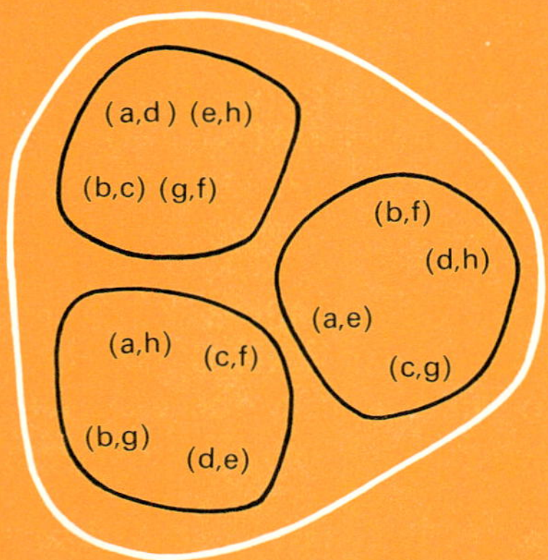

Write all the pairs of equal angles.

Classify them according to their type.

The following game will help to establish the different types of 'equal angles' and, at the same time, open up a further investigation:—

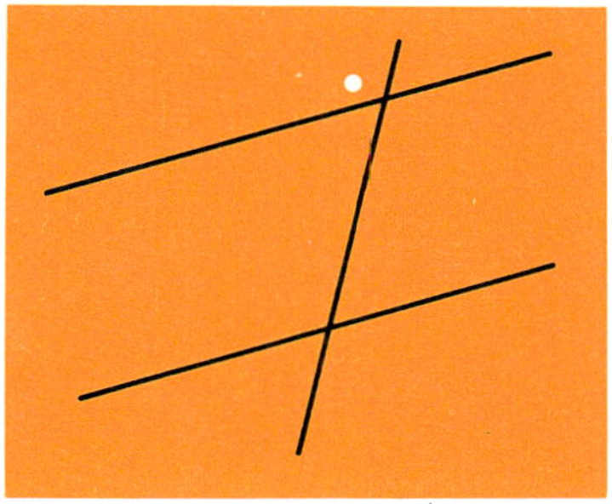

On a new diagram mark one of the angles.

Use symbols V: vertically opposite; C: corresponding; A: alternate. (It is often found that there is a wish to include a=h and b=g as alternate pairs).

Starting at this angle (a) where will I go if I say V?
Where will I go if I say C? A?

The pupils can make a speed game of this with each other—without touching the diagram! Each time starting at the same angle.

What happens if I do V then C then A?
What about V A C?
If I do C and then V where am I?
Yes—at angle 'h'.
How could you get from 'a' to 'h' in one move?
So we could write V C=A.

If pupils are encouraged to do pairs of operations they will soon come across VV or AA or CC and so need to talk about a move called 'stay where you are'. This is the identity operation.

		2nd			
		I	V	C	A
	I	—	—	—	—
1st	V	—	I	A	—
	C	—	—	—	V
	A	A	—	V	—

Take all possible pairs of operations and the single operation that can replace each one.
Make a table of these results.
What do you notice?

Depending on the interest of the pupils a number of patterns will be noticed.

patterns in polygons

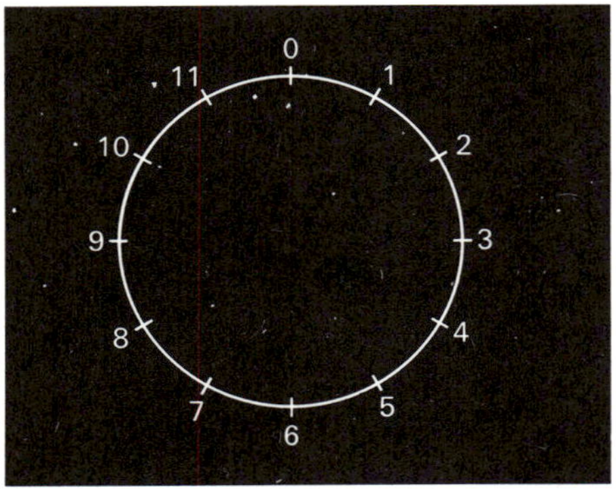

A circle with twelve points equally spaced on the circumference is drawn on the blackboard. (A good supply of duplicated sheets with a similar drawing for students to use is essential.) The points are numbered from 0 to 11.

Can you suggest any rules for joining the points?

—Join the numbers that add up to 11.
—Join to the next one.
—Miss out one.
—Join them all to the centre.
—Join a number to its double.

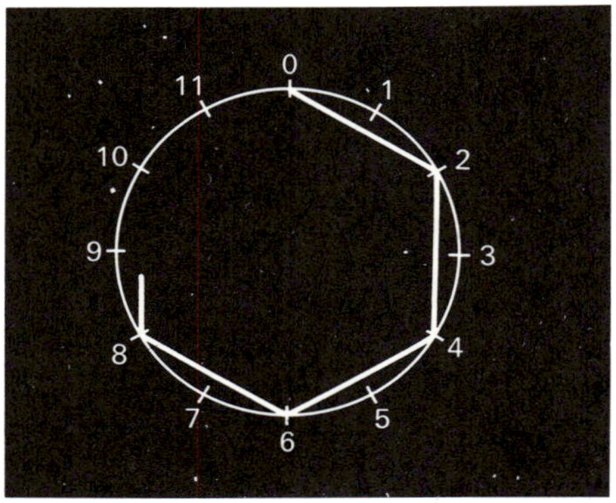

The various suggestions are followed up. Some lead to interesting problems. For example consider the suggestion 'miss out one'.

Will we get back to our starting point?
What shape is formed by the lines?
What happens if we miss out four?

Do we always get back to our starting point?
Can you suggest a number which may not get us back?
For what numbers do we return to our starting point after going around the circle once?
How many times do we go around the circle for other numbers? Why is this?
How many polygons are formed in each case?
What happens if we vary the number of points on the circle?

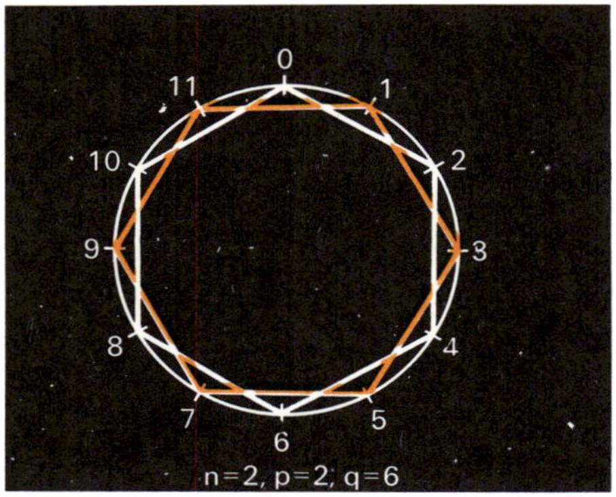

n=2, p=2, q=6

The results for the 12-point circle might be collected into a table:

joining every *n*th point	1	2	3	4	5 . . . *n*
produces *p* polygons	1	2	3	4	1 . . . *p*
q-sided polygons	12	6	4	3	12 . . . *q*

What patterns are observed?

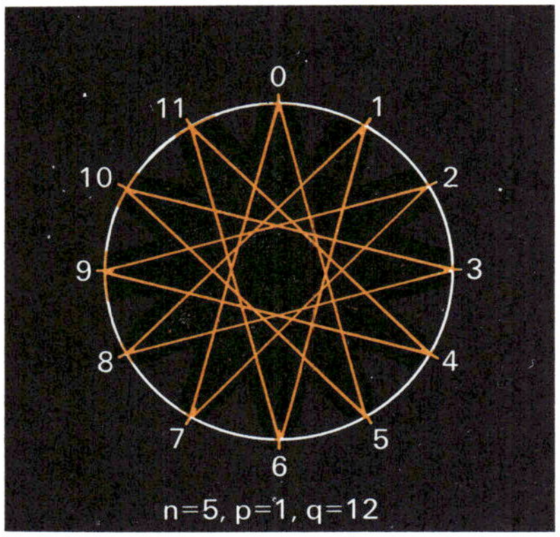

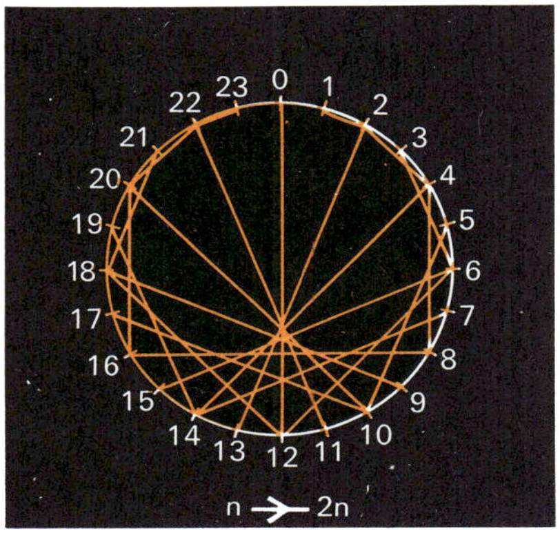

Which *n*th joins give the same figure with reversed direction?

Which *n*th joins give star polygons?

How many derived polygons for a given *n*?

How many star polygons? (Fermat's theorem).

The class might now carry out a similar investigation with different numbers of points on the circumference.

If the suggestion 'join every point to its double' is followed up it will be necessary to consider the 12 points as a number system in mod 12. For example, 2 will represent the class of numbers (2, 14, 26, 38 . . .)
(A circle with 24 points gives a more impressive figure)

To which point will 6 be joined?

To which point will 7 be joined?

The lines form the envelope of a cardioid.

What happens if *n* is joined to 3*n*? *n* to *kn*?

pegboard game

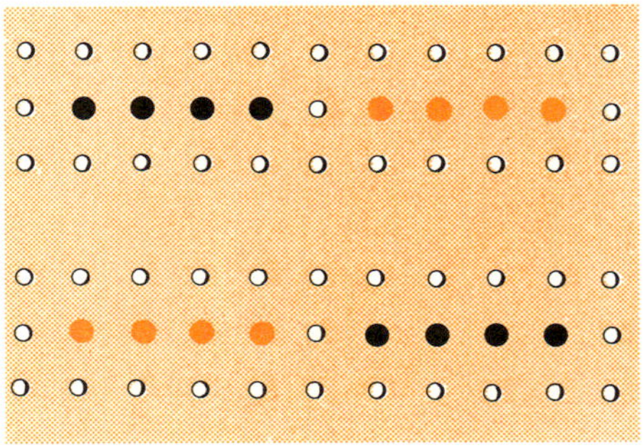

This situation is based on a puzzle due to the French mathematician, Lucas.

Four black pegs and four coloured pegs are set-up on peg-board as shown. The object of the puzzle is to interchange the black and coloured pegs.

Moves may be made according to the following rules:

(a) only the 9 holes (i.e. those occupied by the pegs and the hole between them) may be used.

(b) black pegs may be moved only to the right and coloured pegs only to the left.

(c) a move may be made either by moving a peg into an adjacent hole which is vacant, or by jumping over one peg of the other colour and landing in a hole which is vacant.

The class needs to be given ample time to try this puzzle. Even after some time it is unlikely that many pupils will have solved it. This gives an opportunity to discuss ways of tackling a difficult problem, and, in particular, the possibility of making the problem simpler. One of the suggestions, usually made, to simplify this puzzle, is to reduce the number of pegs.

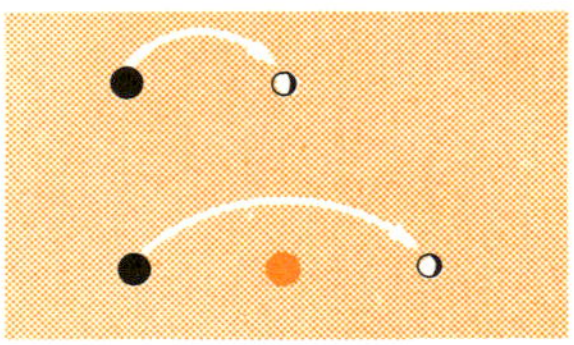

Try the puzzle with only one peg of each colour.
How many moves does it take?
Find some way of recording the moves.

This might be done by recording the order in which colours are moved, or, by recording the moves as slides and jumps.

Increase the number of pegs to two of each colour. How many moves now?

What is the pattern of moves in terms of colours? In terms of slides and jumps?

At this stage can you guess the pattern of moves for 3 pegs of each colour, and the number of moves required?
Try it.
Now solve the original problem of 4 pegs of each colour.

The minimum number of moves (n) required for the number of pegs (p) is a function of p. Can you express p in terms of n?

By studying the patterns of the moves can you prove this function to be correct?

The puzzle might be varied by altering the jumping rules or the initial positions of the pegs. For example, how is the puzzle affected by having two vacant holes between the pegs?

Investigate the game when 3 pegs of each of 3 colours are used; the object being to reverse the positions of the left and right triples with the middle three finally in their initial position.

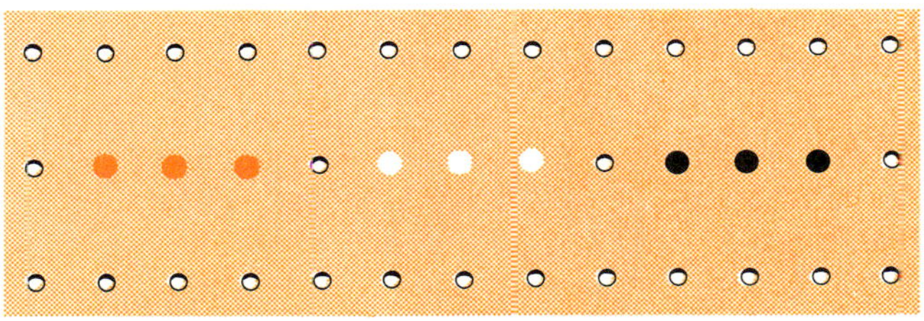

polyominoes

We are familiar with the shape of a domino which is formed by placing two squares together. Polyominoes are simply an extension of this idea into the shapes which can be formed by placing other numbers of squares together in such a way that squares touching have a common edge. If polyominoes are used as classroom material it must be for the mathematical activity they generate rather than the value of the end results. There is no particular value in knowing that there are 35 hexominoes but the activity of finding these can be very rewarding.
See S. Golomb, *Polyominoes*, Allen & Unwin, 1966.

Pupils will require squared paper and scissors. They may find it advantageous to cut out squares and manipulate these instead of drawing shapes on the paper. For demonstration the teacher might use large squares such as polystyrene ceiling tiles.

How many different shapes can be made with 3 squares?
—Can we have shapes like this?
—Is this one all right?
—We shall have to make up some rules.

Various rules could be formulated. If the activity is restricted to drawing on squared paper the rules will be concerned with edges and corners which meet. The work that follows uses the rule that squares that are joined must have a complete edge in common.

How many different shapes can be made from 6 squares?
This problem might be tackled by a group or a whole class of pupils. A record of the shapes found can be kept by sticking the shapes on a large sheet of paper, and numbering them.

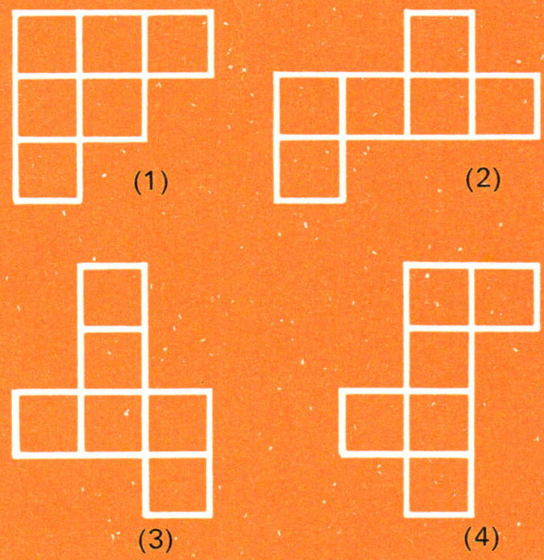

—I've got another one.
—Number 2 is the same as number 4.
—No it isn't.
—Yes it is. If you turn it over and turn it round it is the same.
—I've got another one.
—That's the same as number 29.
—How shall we know when we have them all?

The last remark raises the question; is there a systematic way of finding all the possible shapes? One way is to consider all the squares which can be formed with 5 squares. On each of these place a sixth square in all possible positions and so derive all the shapes for 6 squares. This method although producing several repetitions will provide all the possible shapes.

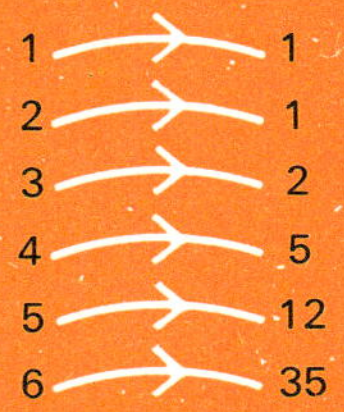

Some further activities:

1. Which hexominoes can be folded to make a cube?
2. Consider the relation between the number of squares and the corresponding number of shapes. Are there any patterns in these ordered pairs?
3. Make the 12 different shapes which can be made from 5 squares. Can you arrange these shapes to form a rectangle?, two separate rectangles?
4. Which shapes can be used to make a tiling?
5. Consider similar problems in three dimensions. For example, how many shapes can be made from a number of cubes?
6. Consider similar problems using shapes other than squares; for example, equilateral triangles, regular hexagons.

probability fairground

Two pupils spend an hour tossing a coin or throwing a dice or choosing a card from a pack. Results are recorded and they have these results to graph, observe and analyse. The work involved in producing the results is often boring and certainly narrow. The starting point that follows provides a way of producing results simultaneously for a variety of activities.

This gives each pupil an opportunity to see a number of different experiments develop. If the pupils analyse the results of an experiment in pairs then the class size will decide the number of experiments to prepare.

A number of sheets of squared paper (1, 2 or 3 cm squares according to need) are prepared giving instructions for the activity and axes drawn ready for use. Some examples of these sheets are shown on the next page.

When the required number of experiments have been prepared they can be spread around the room with the necessary equipment attached to each. It is convenient if the desks are bunched in small groups to allow easy movement around the room.

The class is told to have two or three 'goes' at each experiment, record their results, and then move on.
Pupils have a habit of continuing the experiment after one of the columns is filled so careful observation of the growing graphs is necessary. The experiments should be completed in about 40 minutes.

There are now, perhaps, 18 sets of results already graphed and ready for analysis. (It may be preferred that younger pupils should draw their own graph. In this case they could be asked to make a fair copy, on smaller graph paper, of one of the graphs).

Tossing three coins

Toss three coins onto the table. Record the number of heads uppermost by shading a square in the correct column

Throwing two dice

Throw the two dice onto the table. Add the numbers uppermost. Record the total by shading a square in the correct column.

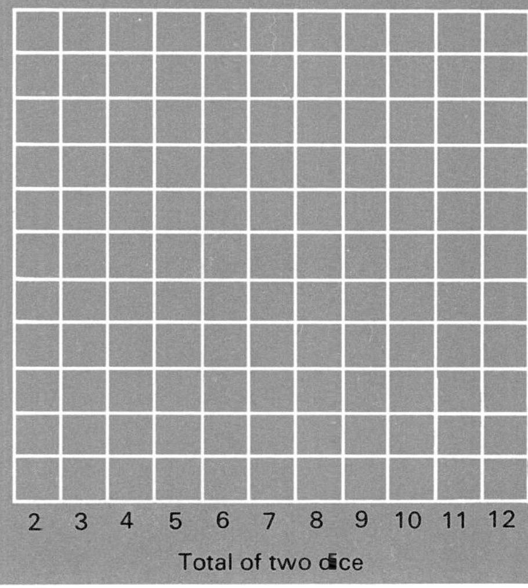

Five coloured cubes

Throw the five cubes onto the table. Record the number of red faces uppermost by shading a square in the correct column

Choosing a card

Choose a card from the pack and record its (a) colour (b) suit (c) court or not.

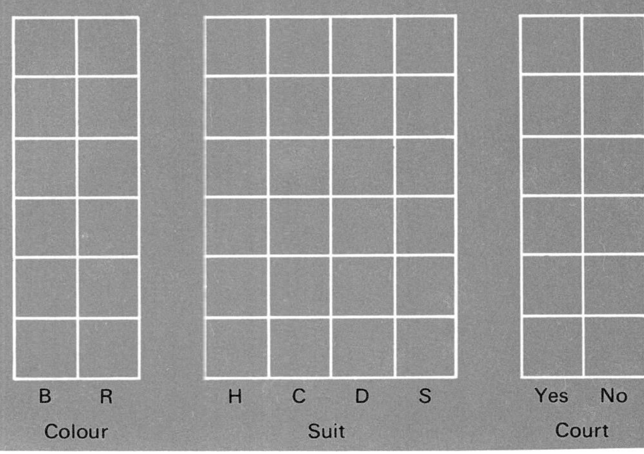

probability fairground
—continued

Before the next lesson each graph can have attached to it a series of questions asking the pupil to make direct observations and to analyse the results. Pairs of pupils could produce a piece of work based on one of the graphs and attempts could be made to link the results between experiments.

To cope with varied abilities in a class it is useful to put only the questions referring to *observations* on the sheet so that further questions can be added, according to the interest and ability of the pupil, during the lessons.

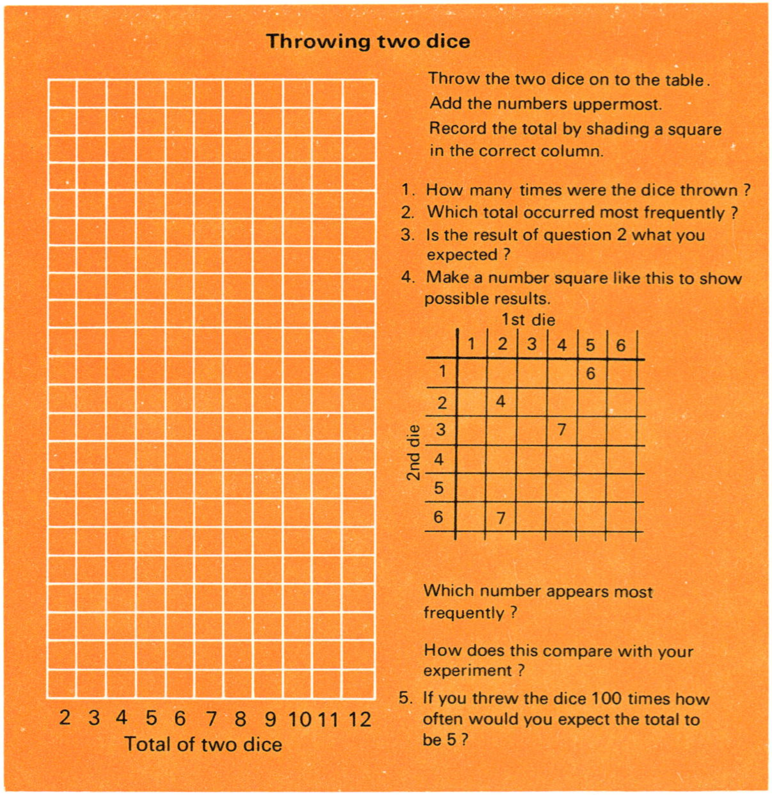

There follows a list of possible experiments with instructions for the pupils.

1. THROWING TWO DICE. Throw the two dice on to the table. Add the numbers uppermost.
   Record the total by shading a square in the correct column.
   N.B. This could be done with any number of dice.
2. TOSSING TWO COINS. Toss the coins on to the table. Record the number of heads uppermost by shading a square in the correct column.
   N.B. This could be done with any number of coins.
3. CHOOSING A CARD. Choose a card from the pack. Record (a) the suit, (b) the colour, (c) whether it is a 'court' card.
   Shade a square for each of the cases, in the correct column.
4. CHOOSING BEADS (PEGS). A bag or sampling bottle need to be filled with several hundred beads of two or three different colours in easy proportions of the total which are known by the teacher, say: 200 red, 300 blue, 100 white.
   Shake the bag. Take five beads. Record the number of red beads by shading a square in the correct column. Replace beads. Shake bag. Repeat.
5. COLOURED CUBES. Five wooden cubes can each have one face coloured red (or two faces).
   Throw the cubes on to the table. Record the number of red faces uppermost by shading a square in the correct column.
6. DECAY DICE. A bag of 100 'decay' dice are produced by Cuisenaire (for address, see p. 232).
   Throw the 100 cubes on to the table. Remove the cubes with a red face uppermost. Record the number of cubes left. Take these cubes and throw them on to the table. Remove the cubes with a red face uppermost. Record the number of cubes left. Take *these* cubes . . . etc.
   Continue this until there are no cubes left. (See also p. 185).

number of throws	0	1	2	3	4	5	6	–
number of cubes left	100	63	31	20	13	–	–	–

quadratics

$x^2 - 5x + 7$

x	$x^2 - 5x + 7$
10	57
7	21
0	7
6	13
3	1
1	3
5	7
2	1
4	3
9	43
8	31

The quadratic expression is written on the board.

Choose a whole number from 0 to 10. Substitute the number for x and calculate the value of the expression.

The results are collected in a table on the blackboard.

What do you notice about the values of the expression? (It might be worthwhile rearranging the table putting the values of x in order).

—all odd numbers.
—lowest value is 1.
—the values get smaller and then start getting bigger.
—a lot of the numbers are prime.
—the differences go 4, 2, 0, 2, 4, 6, 8, . . .
—numbers end in 3, 7 or 1.

By choosing other values for x can you make the value of the expression any lower?

Have you tried negative numbers?
fractions?
zero?

What is the lowest value of the expression?

What is the highest value?

Can you visualize the shape of the graph of this expression?

The graph might be drawn and then other expressions of the form x^2+ax+b investigated in the same way, and the effects of varying a and b observed. Alternatively, quadratic expressions which generate prime numbers make an interesting investigation.

$x^2 - x + 41$

x	$x^2 - x + 41$
0	41
1	41
2	43
3	47
4	53
5	61
.	.
.	.
.	.
.	.

The quadratic expression is written on the board and values of x again substituted.

Observations about the values of the expression will undoubtedly raise the question of prime numbers.

Will this expression always produce a prime number? (Prime numbers are produced for values of x from 0 to 40).

Can you see why it will not produce a prime number when x is 41?

83	84	85	86	87	88	89	90
82	61	62	63	64	65	66	91
81	60	47	48	49	50	67	92
80	59	46	41	42	51	68	93
79	58	45	44	43	52	69	94
78	57	56	55	54	53	70	95
77	76	75	74	73	72	71	96
					99	98	97

If numbers are written in a 'square' spiral commencing at 41 it is seen that the numbers along a diagonal are in fact the values of the expression $x^2 - x + 41$.
Why is this?

In the same way the values of the expression $x^2 - x + 17$ lie along the diagonal of a spiral commencing at 17. This expression also generates many prime numbers.

59	60	61	62	63	64	65	66
58	37	38	39	40	41	42	67
57	36	23	24	25	26	43	68
56	35	22	17	18	27	44	69
55	34	21	20	19	28	45	70
54	33	32	31	30	29	46	71
53	52	51	50	49	48	47	72
					75	74	73

Will the values of all quadratic expressions lie along a diagonal of a spiral?

Which quadratic expressions are good prime number generators?

rotating shapes

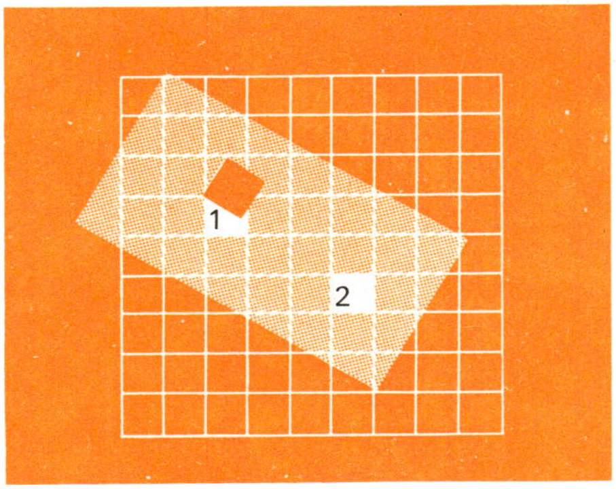

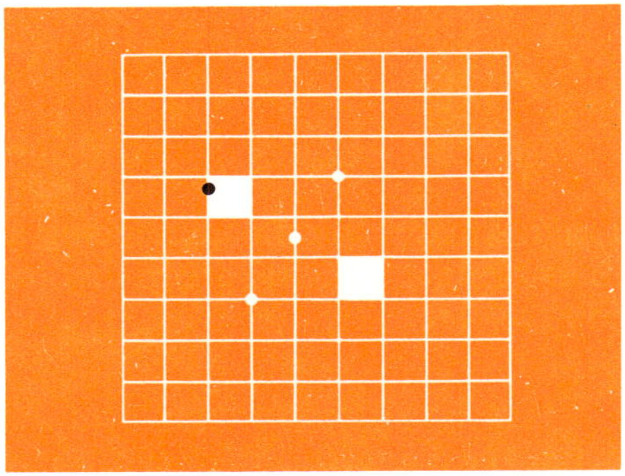

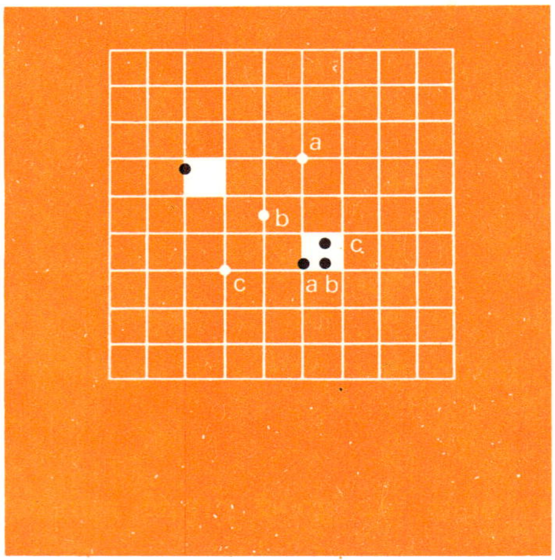

Materials required: some pieces of acetate sheet (see p. 158); square, triangular and hexagonal grid paper.

Shade two of the squares on a square grid.

Place a piece of acetate over the paper and copy one of the squares (keep the acetate square near the centre of your acetate sheet).

Move the acetate square from square 1 on the grid to square 2 on the grid.

What kind of movement is required to do this?

A number of different suggestions will be forthcoming—including: pick up and put on, pick up and turn over, three along and two down, two down and three along, . . .

Discussion will emphasise that there are all kinds of movement but on a particular occasion there may be an interest in working in the plane.

At this stage someone may have 'turned' the acetate to move from square 1 to square 2. If not . . .

Can you find any other ways of moving from square 1 to square 2?

Someone finds a point of rotation.

Are there any other points of rotation?

Any others? . . .

Mark the points of rotation on the square grid.

What do you notice?

Mark one corner of square 1. Copy this on to the acetate square.

What happens to the marked corner if you rotate about point 'a'? point 'b'? point 'c'?

Label the corners of square 2 with an 'a', 'b', 'c'.

How can you get to the unlabelled corner?

Do a combination of rotations. Start with the acetate square over square 1 with the marked corners coinciding. Rotate about 'a' on to square 2, about 'b' back to square 1 and then about 'c' on to square 2 again.

Where is the marked corner now?

To which single move is this equivalent?

Can you get to the unmarked corner now?

Investigate other combinations of three letters (aab, aba, cab, bca, ccc, . . .).

Pairs of rotations will take the square back to itself. Find all possible 'pairs' and find a way of recording this.

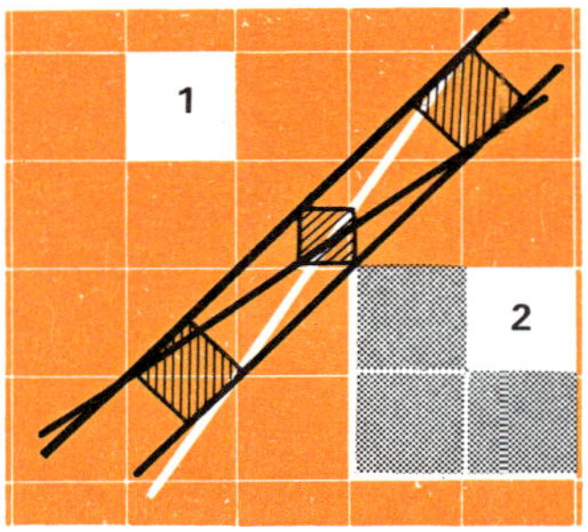

At this stage a number of different questions can be around for different pupils or groups of pupils to pursue.

What happens if the position of square 2 is changed?

One cannot be sure how this will be interpreted but two possibilities are illustrated.

It is interesting enough that the centres of rotation are on a straight line. At a later stage it can be noted that this line is the perpendicular bisector of the line joining the centres of the two squares.

Consider similar questions for two triangles on an equilateral triangle grid and for two hexagons on a hexagonal grid.

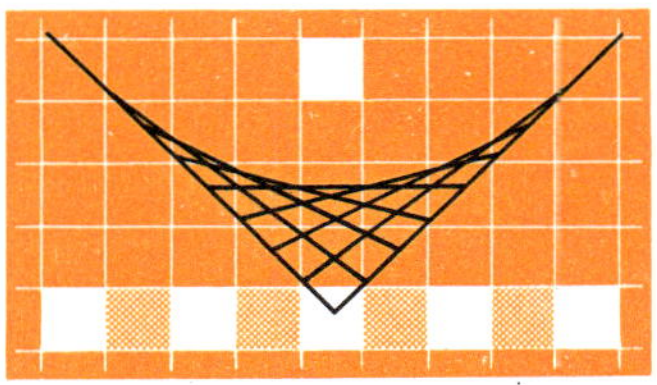

If this activity is carried out on plain paper and the two shapes are differently orientated the number of centres of rotation will link more readily with the symmetry of the shape—4 for the square, 3 for the triangle and 6 for the hexagon.

routes on a cube

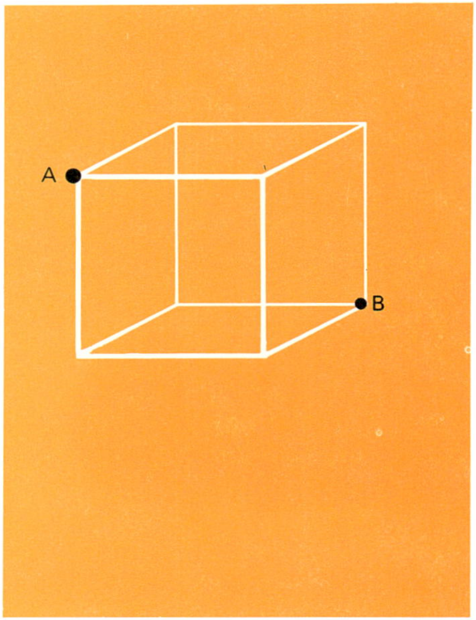

Each pupil has a cube—as large as possible.

Label one corner of the cube A and its 'opposite' corner B. How can you get from A to B?

There are likely to be a number of suggestions including—along edges, across diagonals, along edges and diagonals. . . .

Find another route from A to B, and another, . . .

This finding of more routes will soon provide the need for rules to be made—only three moves, never move backwards, must not go over an edge more than once in a single path, only go round the top once. . . . The rules need to be examined and ambiguity cleared up.

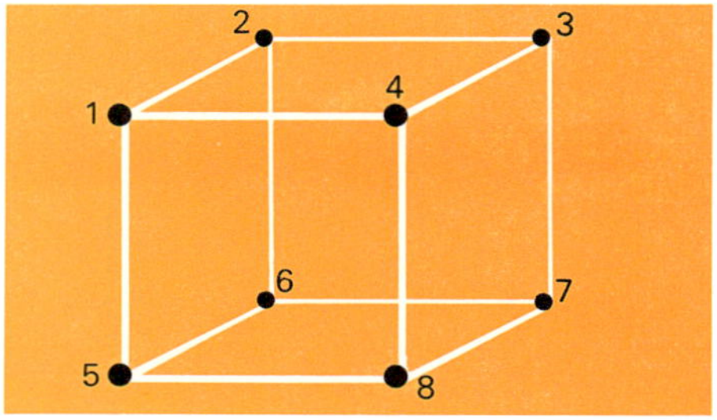

Possible rules are 'shortest path along edges' or 'along only three edges'.

According to the rule, how many routes are there?

The descriptions of routes will show a need for a system of recording. The system might be numbering vertices or labelling directions.

Some of the routes would then be:

	1 2 3 7,	1 4 8 7,	1 5 8 7
or	S F D,	S D F,	F D S.

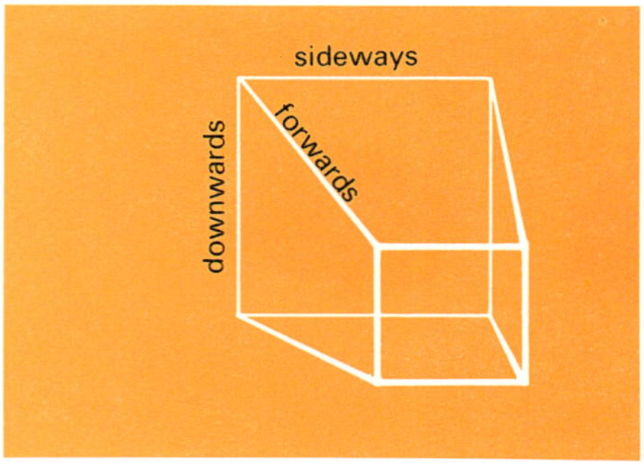

Whichever system is adopted discussion about the way and order the routes were recorded should help to provide some pattern of numbers or letters.

What do you notice about the pattern of the numbers (letters)?

Are there any more?

How do you know?

The shortest route from A to B along edges could be called a '3-edge route'.

How many different '3-edge routes' are there?

How many different '4-edge routes', '5-edge routes', '6-edge routes' . . .?

What do you notice?

What is the longest edge route?

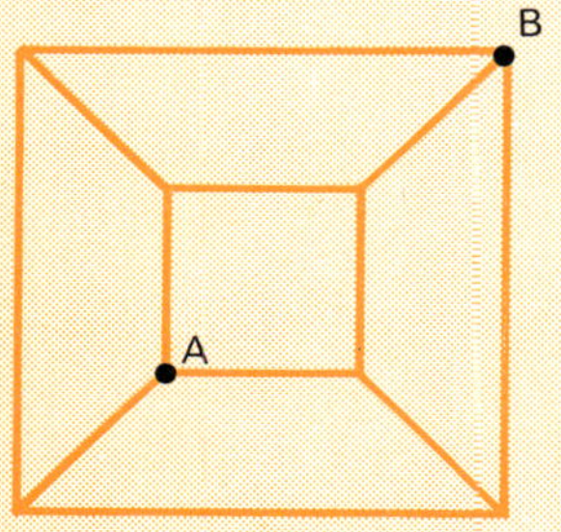

The investigation could be continued or worked in separate groups to find: routes to other vertices, or to investigate the use of different rules.

An attempt to show all possible routes on the cube using different colours leads to chaos quite quickly.

A worksheet of *Schlegel* diagrams, of the cube is a help for this kind of recording and provides a simple 2 dimensional representation of the cube.

Some of the routes mentioned earlier are shown on these Schlegel diagrams.

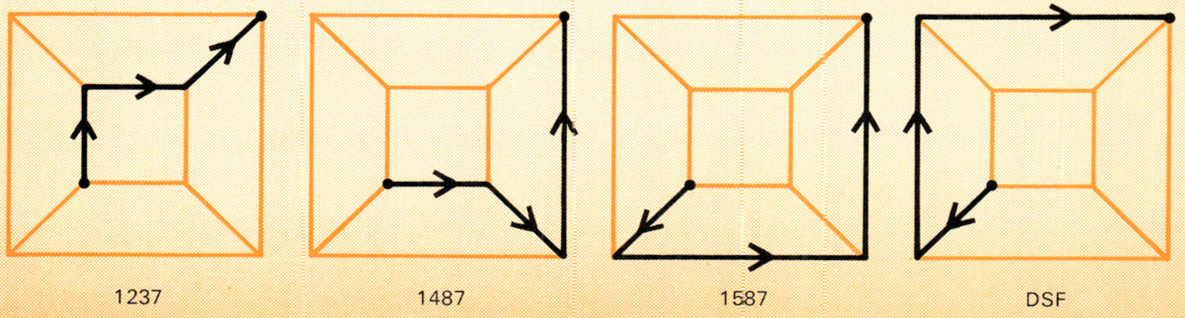

routes on a square lattice

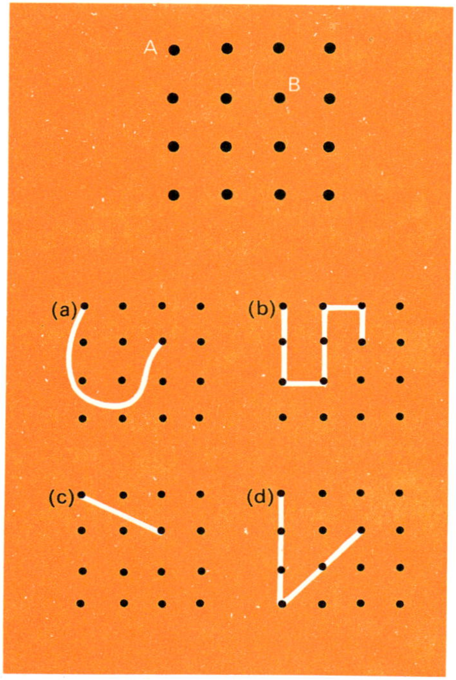

Each pupil has a number of square lattices as shown. On one label the points A and B.

Draw a path, any path you like, which starts at A and ends at B.

Various types of routes will be drawn. Collect a good sample of these on the blackboard and discuss ways in which the routes are the same or different.

In what ways are (a) and (c) the same?
In what ways are they different?
In what ways are (b) and (c) the same?
In what ways are they different?

From this discussion it should be possible to formulate rules for paths which are to be drawn in the work which follows. Here we have chosen rules which are particularly productive, but other rules made by pupils may well lead to interesting work.

We shall consider paths which are drawn along the sides of the 'squares' as, for example, path (b).

Draw one of these paths from A to B.
How long is it?

Some pupils count dots, others count the spaces between the dots. Perhaps the second way is less ambiguous.

Paths of different lengths will have been drawn. Make a list of these lengths on the board.

Can you find some of the missing lengths?
It is not possible to draw paths of certain lengths. Why is this?

What is the longest path?
What is the shortest path?

Investigate the lengths of paths, starting at A, to other points of the lattice.

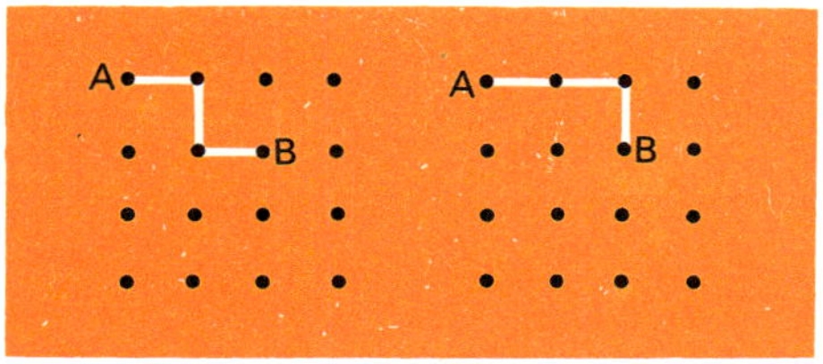

Now we return to the paths from A to B but consider only the shortest.

What is the length of the shortest path from A to B?
Draw a shortest path.
Can you draw a different path from A to B which is the same length?
How many different paths are there?

This number is written against the dot B of the lattice.

Now label another point C.
How many different shortest paths from A to C?
This number is written against the dot C.

This is repeated for other points of the lattice. There is usually disagreement about the number of shortest paths to some of the points and this might well lead to some method of coding. A path can be described by its right and down components; for example:

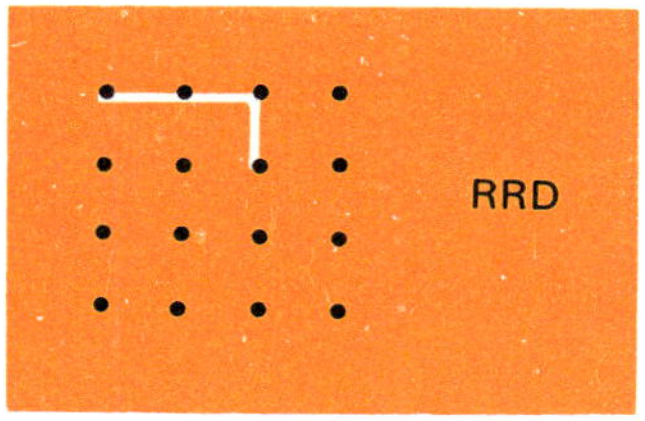

	RIGHT	RIGHT	DOWN
or	R	R	D

is a path from A to B.

It is noticed that all the paths from A to B contain two R's and one D. The permutations of these give the three paths.

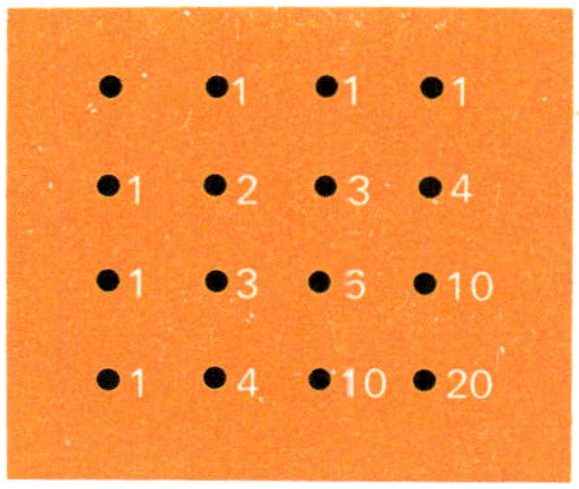

As numbers are written against the points of the lattice various patterns will be noticed and students may recognize Pascal's Triangle.

sampling elastic bands

How strong is an elastic band?

This may well produce such comments as 'depends how big it is' or 'depends on thickness' or 'depends how much weight you put on it'. . . .

What about elastic bands from the same packet which are the same colour and the same size?
Will they all break under the same strain?

Some discussion will be needed to decide on the kind of experiment to be carried out to test how strong an elastic band is.

This might lead to the use of spring balance or to the measuring of the 'band's' length at breaking point. Perhaps both of these results could be recorded for each band.

What kind of recording would be useful?
How could you graph the results?

For bands of the same type a histogram with conveniently grouped data and for bands of different sizes a scattergram to establish any correlation between length and weight (breaking) would prove interesting.

How does your graph help you decide on some uses of elastic bands?

sets

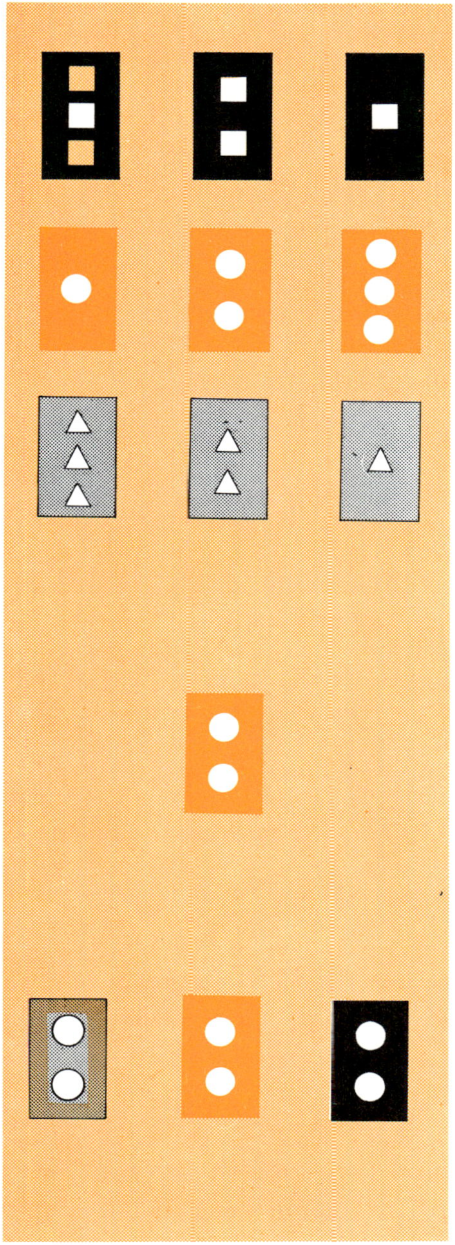

Materials required: 27 cards in three different coloured sets of nine—say 9 yellow, 9 red and 9 blue. Each set of cards containing a card with one circle, a card with two circles, a card with three circles, a card with one square . . . two squares . . . three squares . . . one triangle . . . two triangles . . . three triangles. It is useful to have the cards made so that they can hang on drawing pins, partially pushed into the blackboard. The cards should be large enough to be seen clearly by the whole class. Smaller sets can be made for pupils to use in groups.

To help familiarize the pupils with the cards display all 27 cards for discussion.

Would you like to say something about the cards?

It is soon realized that the relevant properties are colour, shape and number.

How many red cards? Why?
How many cards with circles? Why?
How many cards with three shapes?
. . . . ?

Describe this card (pointing to, say, the red card with two circles).

How much do you need to say about a card so that I know which one you are talking about?

Describe this set of cards. (Placing, say, all the cards with two circles in a different place on the board)
How much do you need to say—how many properties do you need to mention?

This kind of activity needs to continue until the pupils are familiar with minimal descriptions for a card or set of cards.

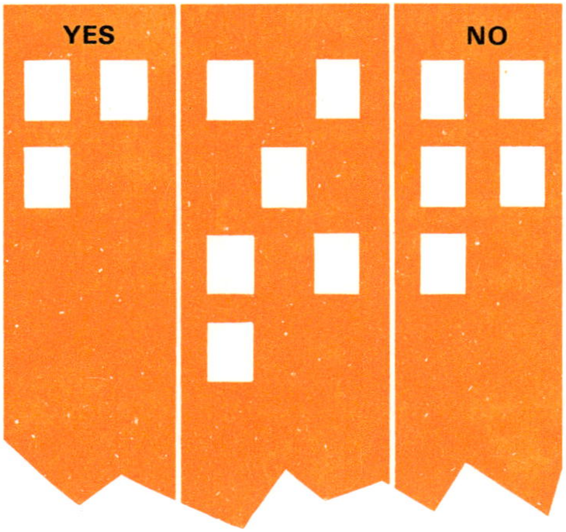

A game: The teacher thinks of a set of cards.

Pupils select one card at a time and teacher says YES if it is in the set and NO if it is not. The 'YES' cards go on one side of the board and the 'NO' cards go on the other. As the game progresses the class is invited to describe the set the teacher has thought of. These descriptions can be tested by taking a card in the 'YES' set and asking:

Does this card fit your description?

It is important that the description of the set can also be used to describe each member of the set.

As the game progresses it is interesting to note the gradually refined descriptions.

R **AND** 3 $(R \cap 3)$

R **OR** 3 $(R \cup 3)$

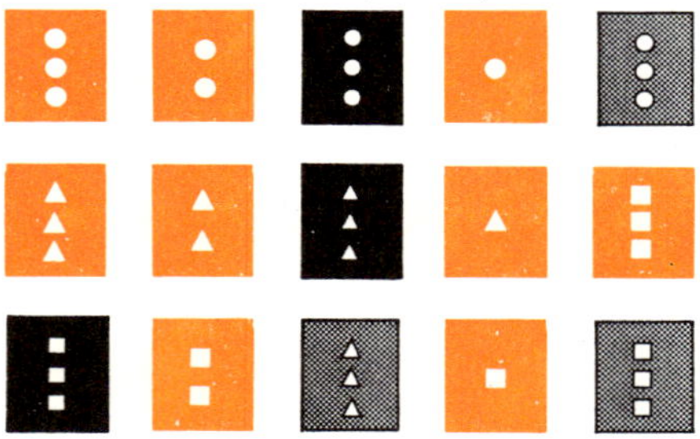

From one set chosen by the teacher the following descriptions came.

—all reds.
—all red squares.
—all red triangles, circles and squares.
—all reds plus blue squares.
—all reds plus blue threes.
—all reds and blue three, yellow three and blue one.
—all reds and all threes.
—everything except blue and yellow ones and twos.
—all red threes.
—everything except blue ones and twos and yellow ones and twos.
—all the threes plus the red ones and twos.

At this stage it can be useful to discuss various ways of writing the above phrases in formal, unambiguous symbols.

If the game is played several times and includes the set R AND 3 as well as the set R OR 3 the distinction between AND and OR will become apparent as attention is directed towards the individual elements of each set.

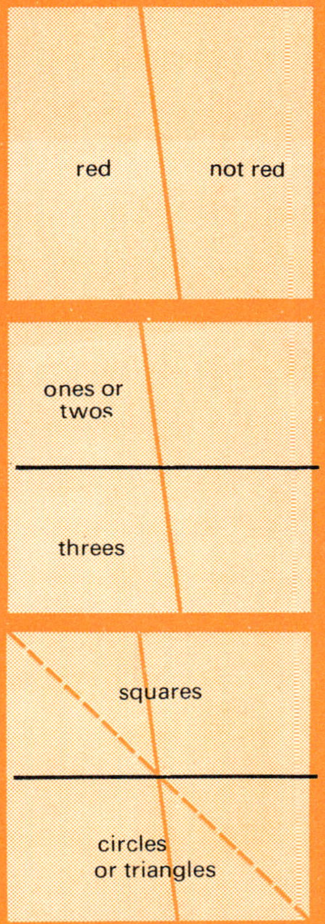

So often, Venn diagrams just appear with circles drawn, no consideration of the universal set and no feel for the partitioning taking place.

The guessing game gives a way of formally classifying the complete set. Although there are three aspects of each card to be taken into consideration the guessing follows a two-fold scheme—each card is or is not a member of a certain chosen set.

A vertical line is drawn on the blackboard in coloured chalk. *The red cards will go to the left, the others to the right.*

Can you do something about collecting the threes?
Draw a horizontal line across the board and put threes below it and others above.

Draw a diagonal line and put the squares of one side and the others on the other side.

Where will I find a set of red cards with not squares but three shapes?
This kind of question will cause some rethinking and some changes in the diagram. This continual changing of the diagram to accommodate any set within connected regions leads to all kinds of methods—some of these are shown.

It is now quite straightforward to introduce set language and more recognizable Venn diagrams.

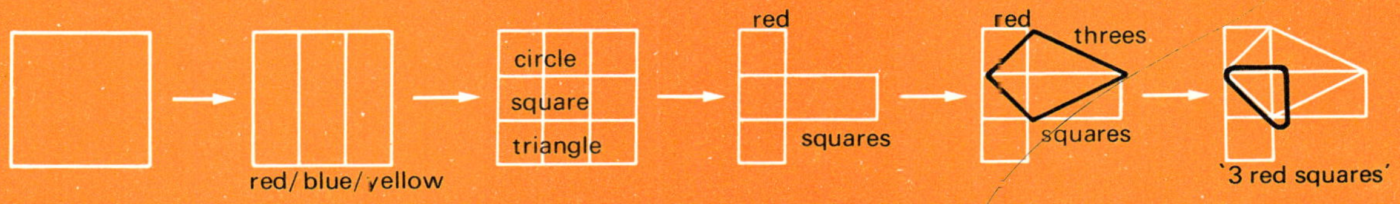

shortest distance

What is the shortest distance between two points?

The reply, 'a straight line', usually assumes that the two points are on a plane surface with no obstacles between them. But what if the two points are:

— on a surface of a sphere, *or*

— two points of a square lattice and distances may be measured only along the sides of the squares, *or*

— two points on a plane surface but the distance must be measured via a given line.

The last option is taken up here.

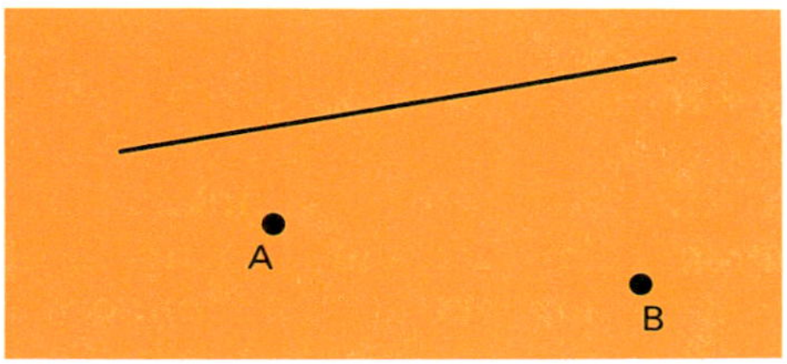

A line is drawn on the blackboard. Someone is invited to mark a point and label it A. Someone else to mark another point on the same side of the line and label it B.

The problem (and it may be described by a suitable story) is to find the shortest route from A to B via the line.

The pupils make various suggestions and each one is examined. In some cases the incorrectness can be seen by varying the positions of A and B; in other cases it is necessary to follow up with constructions. Some of the suggestions are discussed here.

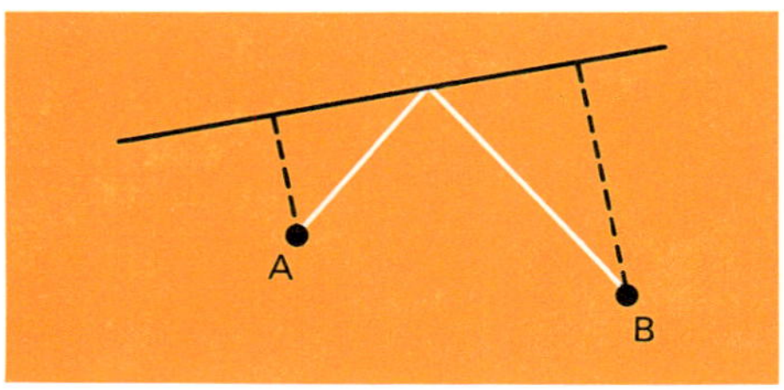

—Mid-way between A and B.

What is meant by mid-way between A and B must be clarified. It usually means draw perpendiculars from A and B to the line and take the mid point between the feet of these. The class discuss this suggestion. Suppose B is very near the line.

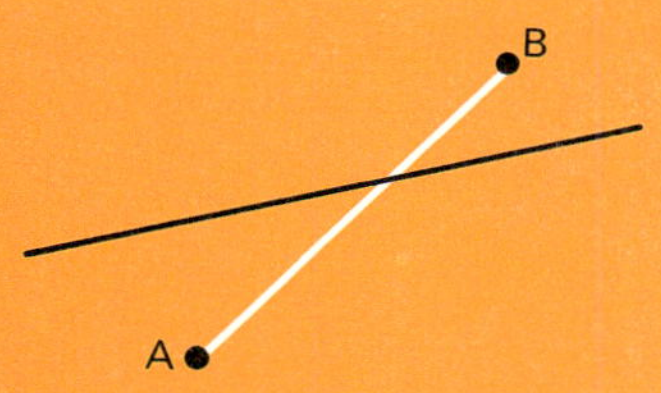

—It would be easy if B was the other side of the line.

This suggestion is far from being irrelevant and may well lead to the correct solution in the end. It might be worth considering each suggestion in the light of B being on the other side of the line.

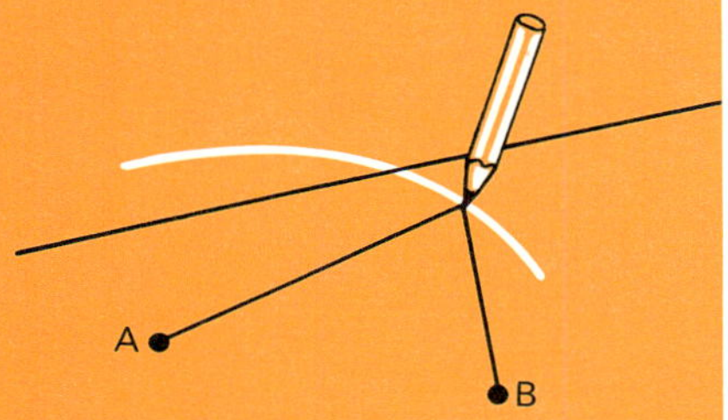

—All the routes are the same length.

Some pupils may object because in extreme cases with B very close to the line the routes are clearly not the same length. However it is worth pursuing further. We start by drawing any route from A to B. To make other routes the same length we fasten a piece of string from A to B which on being stretched taut lies exactly over the route. Now, if we place the point of a pencil in the loop and move it keeping the string taut we shall obtain a locus of points such that any route from A to B via this locus is of the same length. The locus is found to be an ellipse, thus showing that routes via the straight line are not all of the same length.

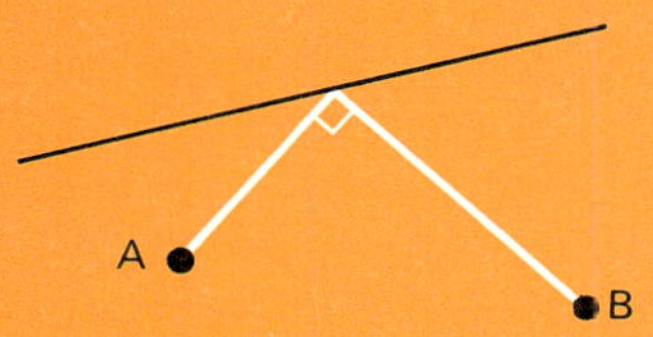

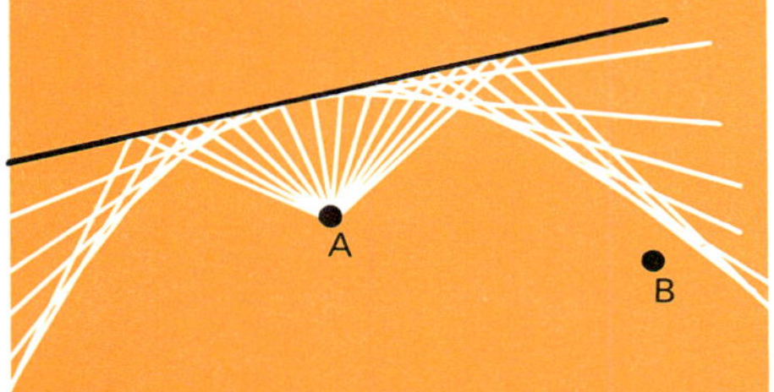

—When the angle formed at the straight line is a right-angle.

This is an interesting suggestion and seems to have possibilities. However, it is soon discovered that for some positions of B it is not possible for AB to subtend a right angle at the line. What positions is it possible for B to occupy so that AB may subtend a right angle? By drawing lines from A to the line and reflecting them at right angles it is found that B must lie in a region bounded by a parabola.

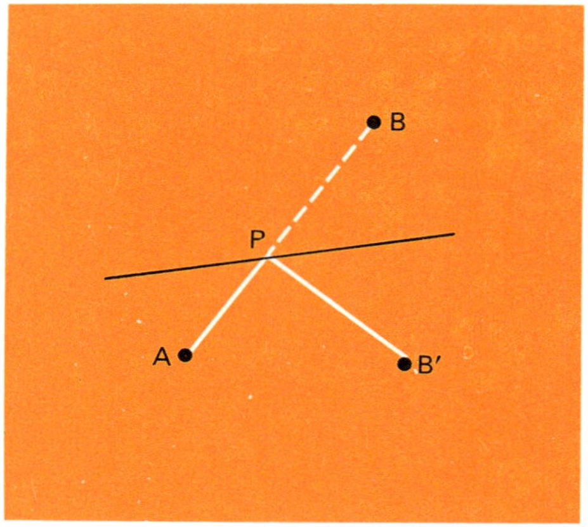

Perhaps the most helpful suggestion has been the second. Consider a route when B is the other side of the line, which crosses the line at P. Now if B is reflected in the line to B′ then APB′ is the shortest route from A to B′. The proof is not difficult.

Other considerations for investigation are:

1. The shortest distance via the line will be the path taken by a ray of light from A reflected by the line towards B. Why?

2. Find other points B_1 B_2 B_3 . . . whose shortest routes from A via the line are the same length as AB (giving a 'circle' by the usual definition).

3. Replace the straight line by a circle or some other conic.

   Given a particular route from A to B via the straight line are there any other routes that are the same length? An intuitive answer would be that there is always one other route. One way of exploring this further might be to construct the locus of all points P of the plane such that PA+PB is constant. The string construction yields ellipses with focuses at A and B. In general these ellipses will cut the line in two points. Discuss the case when the ellipse touches the line.

4. Consider the problem of finding the shortest distance from A to B via a line on a square grid. Here the paths from A can only be along lines of the grid.

six points

Mark six points on the blackboard.

Mark another six somewhere else on the board.

Another . . . .

Are any arrangements the same?

Discussion of this question tends to determine an equivalence of patterns in which distance is ignored.

Make a different arrangement. . . .

After some further work on this the teacher invites someone to join up the points of a particular pattern by straight lines. The aim is to set up a well-defined but implicit equivalence relation through the collinearity of points.

How many lines are there?

—Four.

—Six.

—Eight.

Explain your answers.

Each count is reasonable. In the first place there are 4 lines. But there are 6 segments with points only at their ends. And there are 8 joins in terms of pairs of points connected together. All methods of counting can be used or a single agreement can be sought. The equivalence of patterns will be determined accordingly.

Can you make more straight lines?

How many altogether?

—Eleven.

—Thirteen.

—Fifteen.

What about other arrangements? Find out what number of lines are possible.

Investigate with 7 or more points.

Some work by children is given on p. 44.

shuffles

This lesson is concerned with a particular way of shuffling a pack of cards. To illustrate the way of shuffling, pupils are used as cards and are physically shuffled into new positions. Once the method of shuffling has been established pupils will require cards labelled A, B, C, D, etc. to investigate the situation further.

Four large cards, marked A, B, C and D, are given to four pupils who stand in a column, in order, with A at the front.

The four are now shuffled as follows: A card is taken from the *front* and placed at the back of what is to be a new pile. Next a card is taken from the *back* and placed in front of A. This is continued with B being placed in front of D, and, finally, C being placed in front of B.

What has happened to the order of the cards? (people)

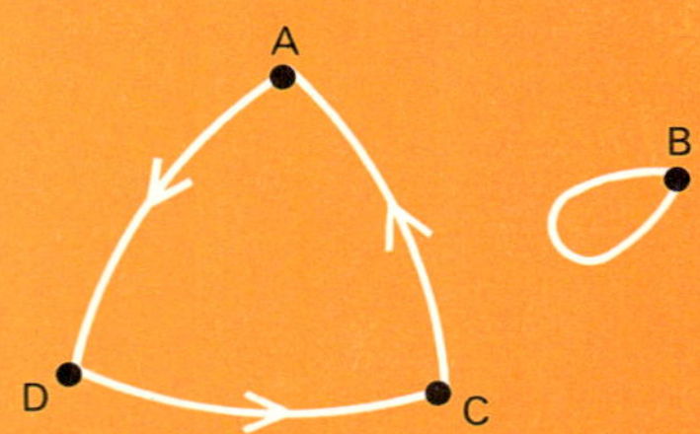

A useful way of showing what has happened is an arrow graph. The arrow shows the card which has moved to the position originally occupied by another card, e.g.:
A →— B shows A has moved to where B was originally.
The graph for the shuffle shows that:
C has moved to where A was
B has remained in its original position
D has moved to where C was
A has moved to where D was.

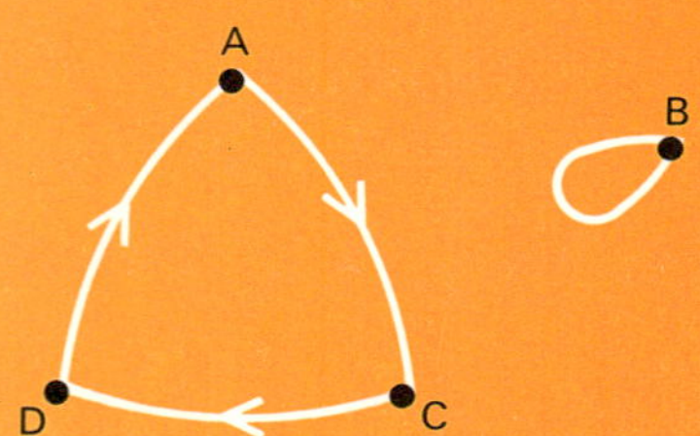

The original order, i.e. A, B, C, D, is now restored. They are now asked to perform a shuffle and, when this has been completed, a further shuffle (i.e. a shuffle followed by a shuffle or s^2).
The arrow graph shows what has happened.

What would happen after a further shuffle (s^3)?
How many shuffles are necessary to restore the original order?

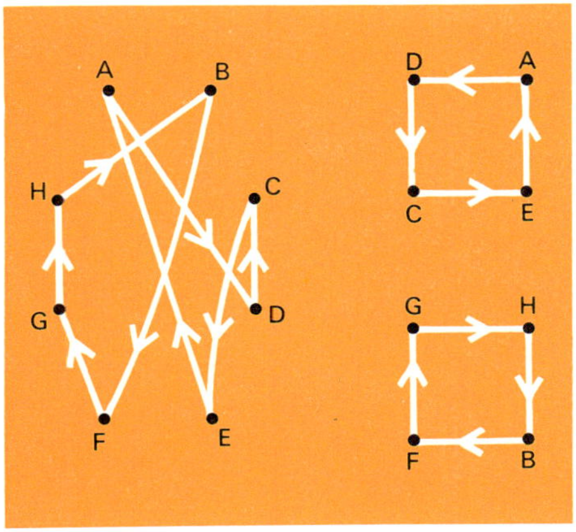

To investigate what happens with other numbers of cards the class might be divided into groups; each group (using smaller cards) to investigate a specified number.

Some graphs can be made simpler if 'disentangled'. For instance, the graph for 8 cards is as shown, but this could be rearranged into a simpler form.

Some relevant questions are:

1. Is it possible, after one shuffle, to forecast the number of shuffles required to restore the original order?
2. The number of shuffles (n) to restore the original order is a function of the number of cards (c). From a table of values of n and c it may be possible to work towards a formula, especially if patterns are noted in values of n for corresponding values of c which are powers of 2; one less than a power of 2, etc.
3. The graph for 6 cards combined with the graph for 4 cards gives the graph for 10 cards. For what other numbers of cards is this possible?

Alternative forms of shuffling are discussed on p. 221.

shunting

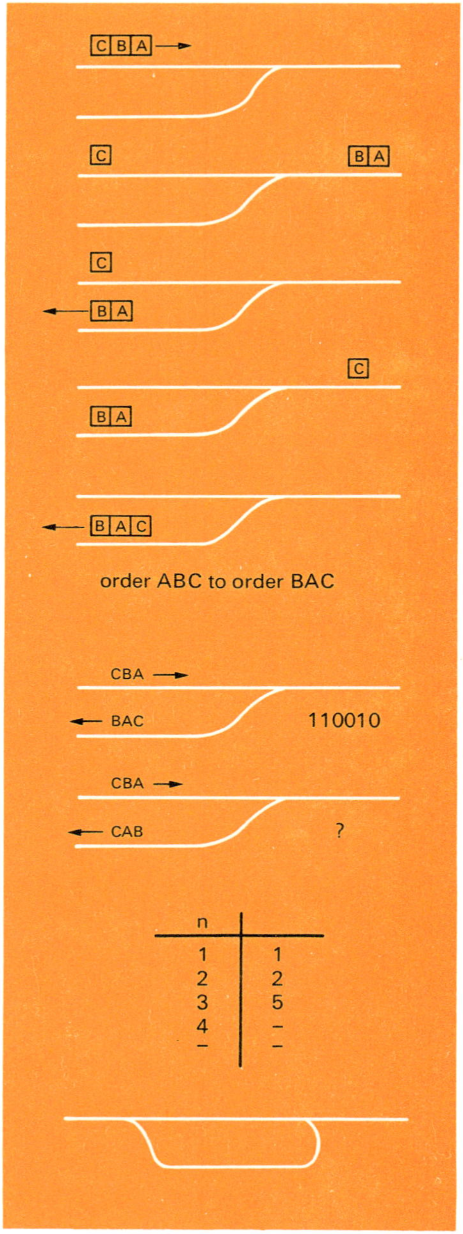

The starting point here is a simple shunting situation. Some trucks come in on one line. Trucks can be shunted one or more at a time into the siding shown and out on to the other line. The trucks are reassembled on the other line.

There are various ways in which the situation can be initiated. Successive stages could be recorded on the blackboard, on a suitably marked piece of pegboard, or by moving pupils themselves.

Try some shunting of a train of 3 trucks.

Can you end up with the original order?

Can you end up with any order you want?

How many operations are required each time?

How can the shunting be recorded?

Discussion may reveal that there are two basic operations—bringing a truck (or pupil) into the siding and taking one out of the siding. These may be coded by two letters or symbols. One class chose I and O for 'in' and 'out' and then gradually changed to numerals 1 and 0.

List the various ways of shunting 3 trucks.

(In the case of 3 trucks it turns out that one order cannot be achieved).

What about 4 trucks?

The list of basic operations on a train of 4 trucks must have four Ins and four Outs.

Can these occur in any order?

Some sequences of I's and O's cannot occur, e.g. IIOOOIOI is not a possible sequence of operations.

How many sequences are possible?

In how many ways can a train of n trucks be reassembled? (The reader may find it interesting to compare this function with the one suggested at the end of p. 168).

Investigate some other shunting arrangements.

sum of a hundred numbers

Sometimes it is useful to ask a question and to let the pupils face the problem with no further help from the teacher. The following description—in their own words—is the result of asking a group of fourth year secondary modern pupils the question:

Can you find the sum of the first 100 numbers?

Solution 1. If you want to find how much a particular number of terms add up to then the number of terms must appear in the answer, i.e. the sum to 7 will have a 7 in it.

(They proceeded as follows:)

2nd Term $=1+2=3=2\times 1\frac{1}{2}$

4th Term $=1+2+3+4=10=4\times 2\frac{1}{2}$

6th Term $=1+2+3+4+5+6=21=6\times 3\frac{1}{2}$

8th Term $=1+2+3+4+5+6+7+8=36=8\times 4\frac{1}{2}$

$$n\left(\frac{1}{2}n+\frac{1}{2}\right)$$

The number of the term is multiplied by half the term plus $\frac{1}{2}$

100th Term $=100\times 50\frac{1}{2}=5050.$

Solution 2.

$1+2+3=(3\times 1\frac{1}{2})+(1\frac{1}{2})$

$1+2+3+4=(4\times 2)+(2)$

$$\left(n\times\frac{n}{2}\right)+\left(\frac{n}{2}\right)$$

$1+2+3+4+5=(5\times 2\frac{1}{2})+(2\frac{1}{2})$

$1+2+3+\ldots+100=(100\times 50)+(50)$

Solution 3.

$1+2+3+4+5+6+7+8+9+10$

First tried (10×1) and the numbers above changed to

$0+1+2+3+4+5+6+7+8+9$

Try 10×2 and the numbers change to

$^{-}1+0+1+2+3+4+5+6+7+8$

Try 10×6 and the numbers change to

$^{-}5+{}^{-}4+{}^{-}3+{}^{-}2+{}^{-}1+0+1+2+3+4$

Try 10×5

$^{-}4+{}^{-}3+{}^{-}2+{}^{-}1+0+1+2+3+4+5$

Seems to be between.

Work with:

$1+2+3+4+5+6+7+8+9+10+11$

$$n\left(\frac{n+1}{2}\right)$$

(11×6) numbers change to

$^{-}5+{}^{-}4+{}^{-}3+{}^{-}2+{}^{-}1+0+1+2+3+4+5$

This works try

$1+2+\ldots+12+13$, etc.

This gives the sum of n terms (for odd n).

When this is checked for even n it works as well.

surfaces and boundaries

Gummed coloured paper, say the 15 cm squares sold in most stationery shops, and scissors should be available.

Cut out a capital letter—perhaps one of your initials.
Talk about your shape.

—it's a C . . . it's got no ends. . . . It's got eleven corners . . . it's coloured on one side . . . it's got a hole in the middle.

It is clearly impossible to forecast what will be said and all sorts of digressions should be allowed. Eventually the teacher conveys by usage and demonstration what he means by the words *side* and *boundary* in this context and asks specifically about these.

My shape has two sides—one coloured, and one gummed.
My shape has one boundary—all round this edge.
Has anyone got a shape with two boundaries?
With three boundaries?

Make a letter H.
Stick any two ends together. Now stick the other two together.
How many boundaries has your shape now?
Has it still got two sides?
Can you make a shape with one side?
How many boundaries has it got?

Some pupils may be familiar with the so-called Moebius strip and may be able to make a one-sided surface from their H by giving twists to the ends that are being stuck together. If not they can be left free to explore the problem. In any case the point is that they should be constructing one-sided surfaces of greater complexity than the usual Moebius strip. Thus the four ends of the letter H may be joined together in various ways.

Various surfaces may now be made from different letters and classified in terms of numbers of sides and boundaries.

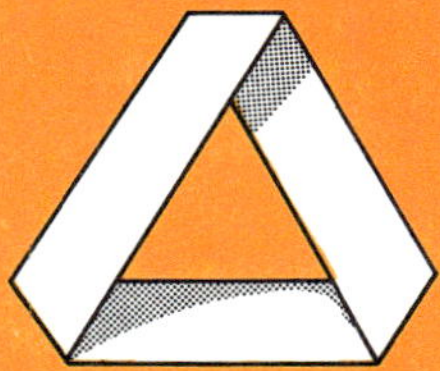
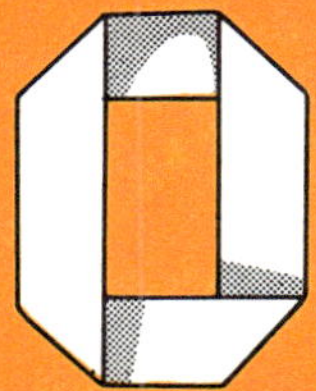

Some further activities:

1. Make an ordinary Moebius band with a narrow strip of paper and flatten out the band so that it will lie flat on a table. Creases will be formed in various ways. Experiment with different ways of doing this.

   Now take some narrow strips of paper or card and form some creased bands by folding over or under a few times and then joining the two ends with glue or sellotape. Pick up the band, uncrease it and see how many twists there are in the closed strip.
   Can you tell how many twists there will be by inspecting the creases of the band in its flattened state?

2. A strip of paper can be knotted to form a regular pentagon. Investigate this construction.

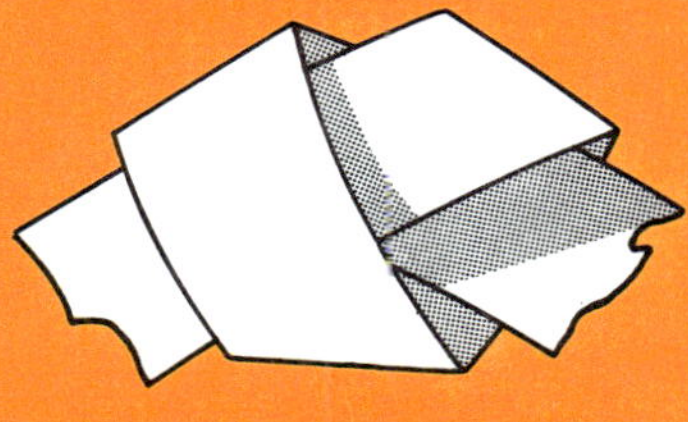
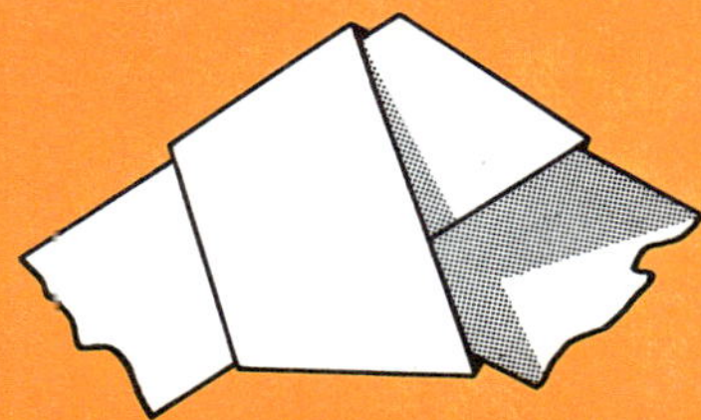

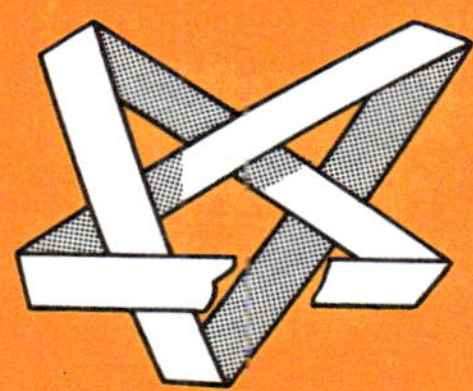

3. Take a piece of string, knot it and then tie the two ends together. Loosen the knot and lay the string flat on a table. Draw a closed curve to represent the knot.
   Conversely, draw some closed curves and see which of them represent knots and which of them represent shapes that can be untwisted, without cutting, to form circles.

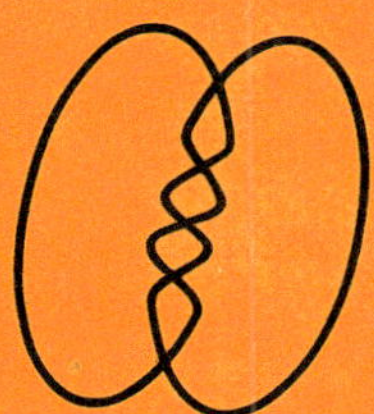

think-a-dot

The starting point is here a plastic toy, *Think-a-dot*, distributed by Electronic Products Ltd, (for address, see p. 232) at £1·10 each including postage, with substantial discounts to schools for sets of a dozen.

You put a ball in one of three holes at the top of the box. There is a series of clicks and a change in the colour of some of the eight spots arranged in a pattern on the face of the box and then the ball emerges from the left or right at the bottom. If you tilt the box to the left or to the right the spots take up certain patterns. A shake of the box seems to give a random initial pattern. Dropping successive balls into the box changes the pattern in what soon appears to be some regular way.

Children can be given the box and left to play. They soon define problems, seek solutions and find patterns.

The notations for describing results vary even more than the problems themselves. In any but the most superficial work with the Think-a-dot some method of recording seems to be helpful. The need for a notation is felt throughout the work. The diagram shows one way of recording the routes a ball can take. Are the problems that a person can consider with the machine, affected by his ability to solve the notational difficulty? As soon as notation is found, it becomes part of the situation, and may itself suggest new lines of thought: one wants to start finding specific sequences or seeing how long a hole will stay blue, or if it always has an unfair distribution of blues and yellows in its sequence. Different recording methods spotlight different results.

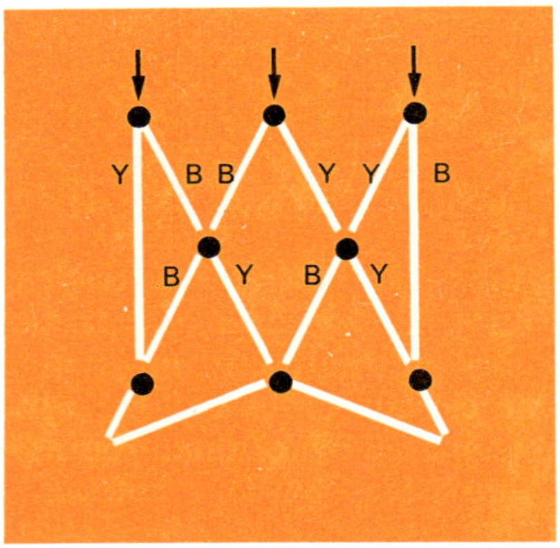

Can you make all the spots blue? Yellow?

What is the least number of moves in which you can get the spots blue from, say, the pattern obtained when the box is tilted to the left?

What routes does the ball take?
And from which side it will emerge?

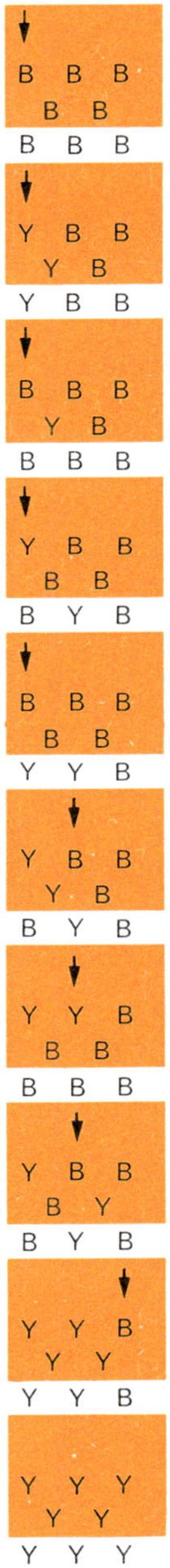

Some further issues:

1. State conditions under which the middle spot in the bottom row will change colour when the ball is dropped into each of the three holes. Describe the changes in the middle row.

2. In how many moves can you change the pattern in which all spots are blue to one in which all spots are yellow? A pattern may be said to be dualised when all its spots change colour. Study those sequences of moves which dualise a pattern.

3. Invent a game for two or more players.
   Can you make a mathematical model of the game?

4. The effect of dropping a ball in the left-hand hole is to change the colour of the left spot in the top row. If the two colours are denoted by 0 and 1, the effect may be described as changing x to $x+1$ where the addition is boolean. Can you characterize the other colour changes using boolean algebra?

5. The initial choice of a hole may be indicated by L, M or R. Does the sequence LR always have the same result as RL? Can you simplify the sequence LMRLRRRM?

6. It is soon found that repeated dropping of a ball in the right-hand hole restores the original pattern after 8 repetitions. This cyclic nature of the operation may be described by an equation, e.g. $R^8=1$.
   Express other cyclic properties in this way.
   Do the elements L, M and R generate a group?
   Investigate the algebra of these symbols.

7. Each spot can be in one of two states, blue or yellow. Thus there could be 2^8 or 256 possible patterns. Starting from a particular pattern can all these be generated by a sequence of moves?
   Characterize any patterns which cannot be generated.

8. Make up some problems for further investigation.

tournaments

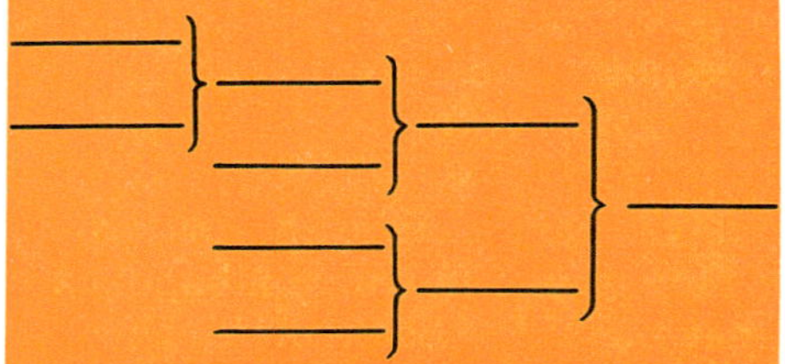

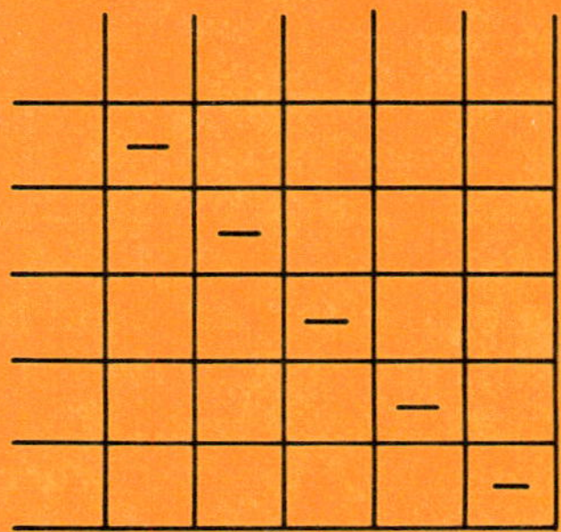

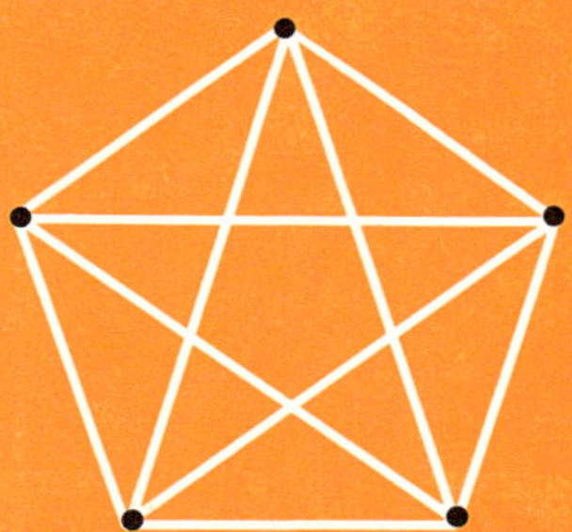

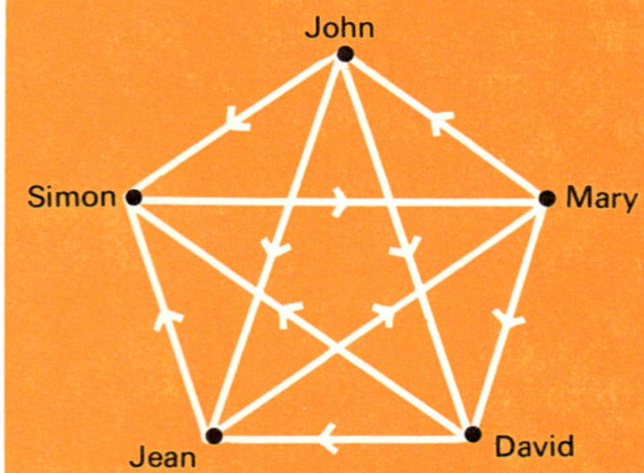

How can we organize a tennis tournament for 5 players? One possibility is a 'knock-out' tournament which in this case has to give some players a bye into the second round.

Another possibility is the *complete* tournament in which each player has a match with every other player.

We take a complete tournament as a starting point.

How many matches are played altogether in a complete tournament for 5 players?

—20, because each player plays 4 others.

—No, 10, because that way counts every thing twice.

—10, because its 4 plus 3 plus 2 plus 1.

How can the tournament be recorded?

This question may arise naturally from the previous discussion. Suggestions include a five-by-five table or a diagram of 5 points with all possible joins. Here we follow up the second suggestion. On this diagram the results of the matches can be indicated by arrows. An arrow from A to B means that player A wins the match with player B. Someone is now invited to put in an arrow on each line at random. The diagram—or graph—becomes directed (cf. p. 164).

What are the results of this tournament?

—John is the winner.

—But Mary beats him.

—But no-one else wins three matches.

—Simon only wins one.

—Three people draw—no, Mary is the best of these because she beat the winner.

Make up your own tournaments by putting in arrows in various ways.

Make up a 'fair' system for deciding on the final order of the players.

Pass your tournament to someone else and see if they give the same final order.

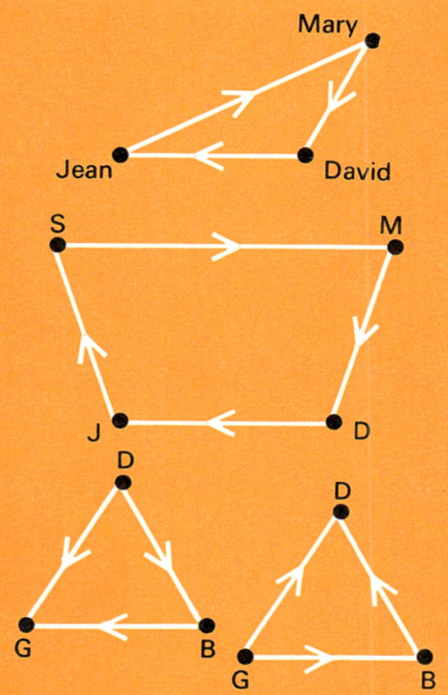

number of players	possible tournaments
2	1
3	2
4	4
5	–

number of wins	greatest number with these wins
1	3
2	5
3	7
4	–
–	–

Some further issues:

1. In the example given, Jean, Mary and David formed a 'cycle' of 3 players each winning against one other.

   Is there a cycle of 4? Of 5?
   Is there a path from one point of a directed graph passing through every other point?

2. How many different results could there have been in a tournament for 5 players?

3. Dorothy, George and Bill played two complete tournaments with the results shown. Dorothy won the first tournament and George the second. But apart from the names both tournaments have the same pattern. How many essentially different tournaments can arise with 5 players where the individual labelling of the players is not taken into account?

4. In a complete tournament for 5 players, 2 players win 3 matches, 1 wins 2 matches and the remaining 2 win only 1. Construct the diagram of the tournament.
   Is there another with the same scores?

5. In a complete tournament for 10 players can there be 8 players who win exactly 3 matches?
   In any complete tournament what are the limits on the number of players who win 3 matches, 4 matches, 5 matches . . . and so on.

6. Consider the option of recording the results of the tournament by a table. The elements of this table might form a matrix that is related to the associated graph.
   Investigate further.

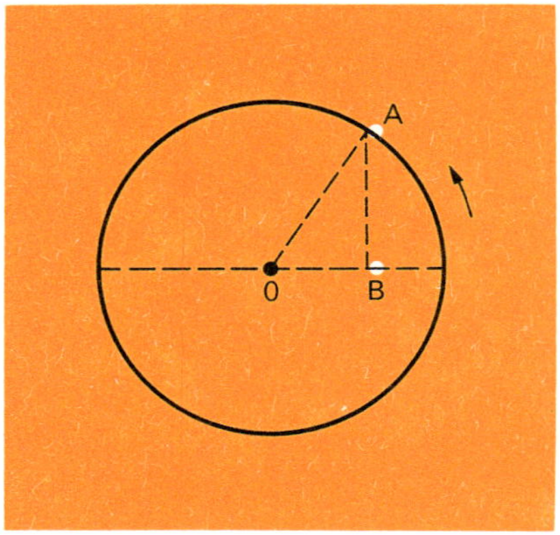

The filmloop *Simple Harmonic Motion* (obtainable from Macmillan, see p. 232) shows the circular motion of a point and the corresponding linear motion of a second point along the horizontal diameter. The 'diameter' point moves in such a way as to be vertically below or above the 'circle' point.

The loop is shown and the pupils invited to comment on what they have seen. The points are given labels, say A and B.

How does the point B move?
How does the point A move?
How can we describe the position of A?

Suggestions might include a co-ordinate system or reference to angles. It is useful to use angular measurement, giving the angle between the rotating line OA (O being the centre of the circle) and the diameter. A diagram on the blackboard is useful at this stage.

Where is A when the angle is 90°?
Where is A when the angle is 150°?
Where is B when A is in this position?

Mark a position of B.

Where is A when B is in this position?
Where is A when B is at an end of the diameter?

Each pupil draws a circle of unit radius (say 5 cm). On it are marked positions of A and the corresponding positions of B. The corresponding A's and B's are joined by straight lines.

What happens to the line AB as A moves around the circle?
When is AB longest? At which angles?
When is AB shortest? At which angles?
Estimate the angles for which AB is about half the length of the radius.

At some stage this line AB is given the name 'sine line'.
A graph showing the relation between the angle and the length of AB gives the sine curve.

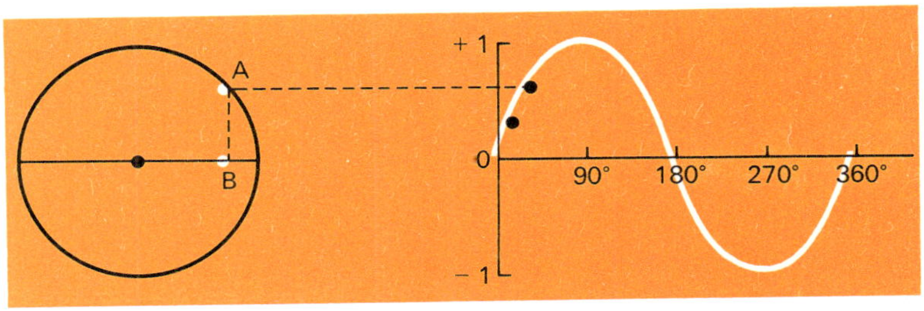

A further showing of the film and consideration of the length of OB in a similar way leads to a graph of the 'cosine line'.

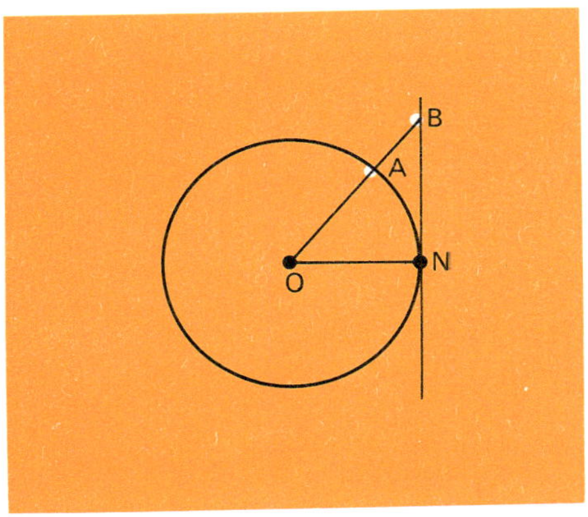

An extension of the theme has been carried out by a group of pupils who scripted and made a film on the tangent line. The resulting filmloop is available from A.T.M. Filmloops—for address, see p. 232.

Where is A when the angle is 90°?
Where is B when A is in that position?
Where is A when B is at N?

Draw a graph to show the relation between the angle and the length of the 'tangent line' BN.

III Materials for mathematics

III Materials for mathematics

introduction

A baby grabs a rattle and throws it out of the cot on to the floor. It is picked up and returned to the cot. The baby throws it out again. He gurgles with apparent delight. An infant plays with bricks, making various arrangements and disturbing them. At night he may rehearse counting, drowsily aware, somehow, of the mystery of an inexorable sequence that seems to have no end. One child bounces a ball against the wall and catches it. Another skips to complicated rhythms. Another plays other games like noughts-and-crosses. He plays with animals, with toys, with dressing-up clothes, with other people, with anything that lies to hand. He plays and plays. . . .

This play has little to do with the strenuous competitions of adult games. In the sense we are using, few adults, sad to relate, can play. What is the point of play? It seems to offer an important challenge to the player and we suggest that in many cases this challenge takes the form of an absorbing curiosity about what happens when one creates a universe. 'What happens if . . .?' Like fledgling gods we can make it happen and find out.

But this creative paradise is soon disturbed because it is even more omnipotent to withhold the happening and pursue the *what-if* in the mind. Here lies the power, and, at the same time, the poverty of the intellect. Now, mathematics is virtual action. As such, it may enable a mastery of the universe. But it also runs the risk of causing a serious regression from reality. We suspect that many of the emotional difficulties incurred in the learning of mathematics arise through too strenuous and premature a pursuit of the *what-if* in the mind, without adequate experience grounded in perception and action.

Mathematics at the secondary level relies heavily on a stock of images and actions on which the mind can operate. It is an important task for the teacher to see that this store is acquired and constantly replenished. So we believe that there should be plenty of material to play with in the classroom.

Everyone pays lip-service to the notion that there should be material for making mathematics in the classroom. But even if some is bought out of the little money available, it often ends up at the back of a cupboard and is not used as a regular feature of a course. Significantly enough, there are many alternative names for such material—apparatus, teaching aids or equipment, audio-visual aids and so on. In such cases the material may be intended to aid exposition of a single topic. When the teacher realizes that, for various reasons, his pupils have once more failed to grasp what he has been telling them, it is quite natural that he will be cautious about using the equipment again or buying more like it. But this is certainly not what we have in mind. We are concerned mainly with material that is 'multivalent' in the sense that it can be used in many ways.

In this section we give an account of some specifically mathematical directions that may be taken with various materials. It is vitally important that children be permitted to play freely and that the teacher be able to commit to the mathematics that they may wish to make. It is equally important that the children be able to commit to the mathematics that the teacher may wish to make. In either case it is helpful if the teacher is aware of some of the directions that can be taken. We would emphasize again that there is nothing prescriptive about our suggestions. They arise from work that has been developed with some children at sometime. We hope the reader will work through some of the problems himself—the questions are mainly addressed to him—and that further work will arise.

We have tried to give some current prices of the materials we describe. Clearly these prices can only indicate the order of cost and cannot be relied on when ordering items. In no sense are we recommending a 'best buy' though we have not mentioned any material or resource that we have not found useful. Some of the materials are available from ordinary stationers or toyshops, some from the major educational suppliers and some from particular manufacturers. We have indicated such sources where possible and a list of their addresses, with others, will be found on p. 232.

In some cases we have given only a very brief account indeed of the activities that can arise with a particular material. This is usually because there are excellent full accounts elsewhere. In these cases, we have more or less just listed the material with some sample activities as an indication that we think it is an important one.

We think it would be helpful to pay some careful attention to storage. A large stock of material inevitably demands good shelf and cupboard space, and a stockroom. It may be worth having a good selection of stout cardboard boxes, such as those that are used in many infant schools. We have found it convenient to use shoeboxes which can often be collected free from shoeshops. Such boxes make it just about possible to move material around, though the more it is used, the more important it is to have specific mathematics rooms, preferably close together.

Finally, a note of caution. There is more to teaching than the provision of appropriate materials. The widely recommended mathematics 'workshop' can be successful in solving certain organisational problems of class teaching. But such success may be misleading, especially where it seems to produce a sort of abdication by the teacher from the dialogue the material can set up. It is difficult to clarify this point in abstract: we can only stress that mathematics is not in material—it is in the mind.

acetate sheet

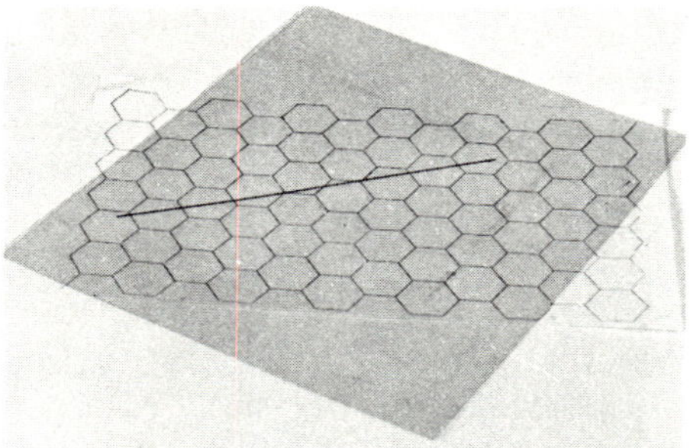

Acetate sheet is a transparent material available from most large stationery shops and printers. It can be obtained in rolls or in single flat pieces and in many thicknesses, e.g. from Ian's Ltd (see p. 232). A recommended thickness is 0·25 mm. Cleaned X-ray sheet, which has a slight blue tint, is quite suitable for the work described below.

A special pencil or crayon is required to make marks on acetate. Chinagraph pencils at about 6p each are generally available; their mark is easily cleaned with soft tissue. Felt pens make clear and attractive marks, but when dry these are more difficult to remove. Marks made with water-based inks can be removed with a damp cloth; spirit-based inks need methylated spirit.

The nature of the crayon or felt pen tip makes it difficult to work with precision, and so most of the situations described are simple and do not require great accuracy.

Draw a line on paper about 10 cm long. Copy this line on to a piece of acetate sheet.

Move the acetate sheet over the top of your paper and stop to show a position of the two lines. Record.

Move again. Record.

If each child is asked to record ten or so different positions for his two lines, and is then asked to join the class in building up a record on the blackboard of different patterns, this should provide a starting point for some discussion about parallels, verticals, horizontals, perpendiculars, angles, intersections and so on.

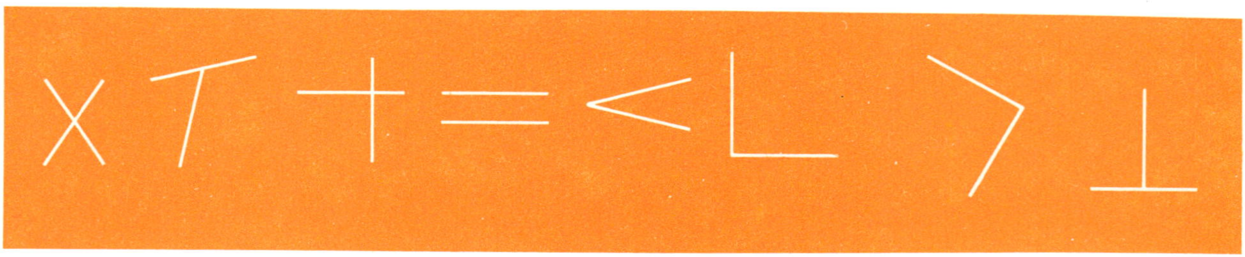

What do children mean when they talk about 'is the same as', 'is different than'?

The elements involved in the previous activity were lines—a line on acetate and a line on paper. Call this (L, L). Other simple yet fruitful situations, with points (P), lines (L) or circles (C), will be referred to in this way. e.g. (P, L)—point on acetate, line on paper, (L^2, C)—2 lines on acetate, circle on paper.

The reader is strongly advised to manipulate the actual material while reading the following notes.

(P, L)

Move acetate, keeping the point on one end of the line.
Mark positions of the other end of the line on the acetate.
Discuss findings.

What happens if the point is not on the end of the line but over some other part of the line? What emerges from marking both ends of the line after each move?

Consider some regular movement with the line marked off in about twenty equal divisions and the acetate rotating as the point moves along the line.

(C, L)

Make the line a chord of the circle. Mark the mid-point of the line. Move the acetate so that the line remains a chord of the circle. Mark the mid-point of the line after each move. Copy the line each time you move the acetate.
Vary the length of the line.

acetate sheet—continued

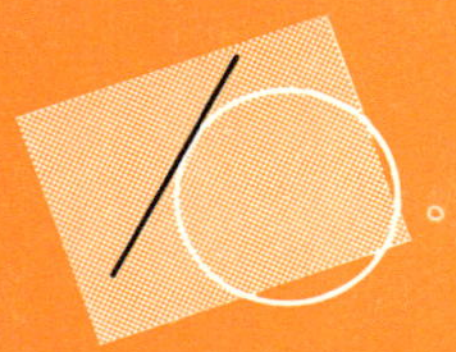

(L, C)

Move the line over the circle to make chords, diameters, tangents.

If the centre of the circle is marked then the 'angle at centre' can be discussed as the chord shortens.

(C, C)

Move the acetate circle over the paper circle.

Consider circles of different sizes, same sizes—record different possible positions of two circles.

From a number of different intersecting circles draw lines joining centres, joining intersections, joining centres to intersections (constructions of perpendicular bisectors, perpendicular to a line from a point, rhombuses, kites). See p. 96.

Draw tangents to circles in different positions. Alternatively, consider (C^2, L) where the two circles can be of different sizes or not, and drawn in many different positions. (The two circles could be each drawn on a different sheet of acetate.)

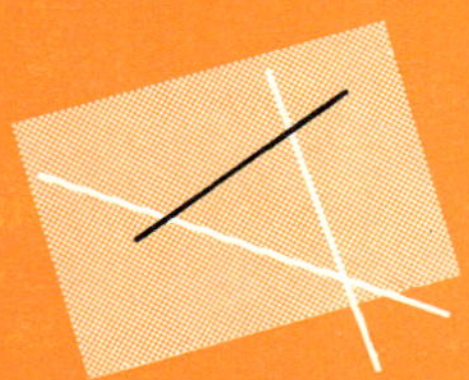

(L, L^2)

The case when the lines on the paper are parallel prompts an intuitive appreciation of some angle properties which can be discussed. See p. 112.

If the lines on the paper intersect then different triangles emerge. Note the changes in the two angles of the triangle that vary.

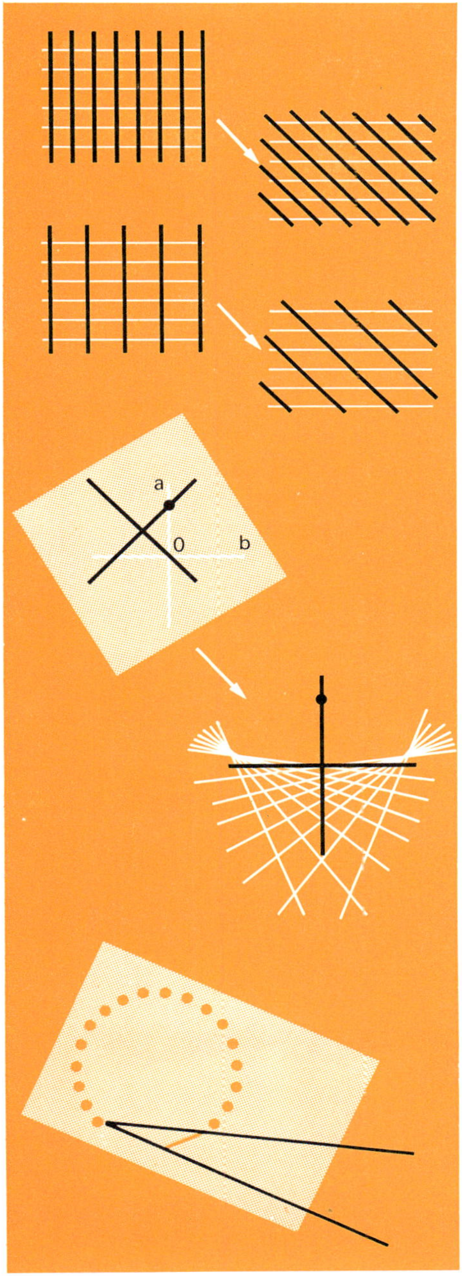

(L^2, L^2)

Consider the case where the lines on acetate are a copy of the lines on the paper. Movement of the acetate produces various quadrilaterals. Where the lines on each are parallel and the same distance apart the movement yields squares and rhombuses. Where the lines on each are parallel but not the same distance apart the movement yields rectangles and parallelograms.

If each of these is extended to (L^n, L^n) then tilings of these shapes emerge.

Consider the case where the lines on paper and acetate each intersect at right angles.

Label the paper lines *a* and *b* and the intersection *o*. Mark a point on one of the lines on the acetate. Place the acetate so that the point on the acetate is on line *a* and the unmarked line on the acetate goes through intersection *o*. Copy the line *b* on the acetate. Move the acetate keeping the point on line *a* and the line passing through *o*. Copy the line *b*. Continue this. Consider the resulting envelope of lines.

What happens when the lines *a* and *b* are not at right angles?

(L, L^2)

Draw the lines on paper so that they are about 10 cm long intersecting at an angle of about 40°. Draw the line on acetate about 4 cm long (near the centre of your sheet).

Place the acetate over the paper so that the ends of the 4 cm line are touching the two intersecting lines (see diagram). Mark the point of intersection of the paper lines. Continue moving the acetate so that the ends of the short line touch the two intersecting lines, plotting the point of intersection after each move.

How is the circle completed?

Consider intersecting lines at different angles to each other, in particular at 90° to each other.

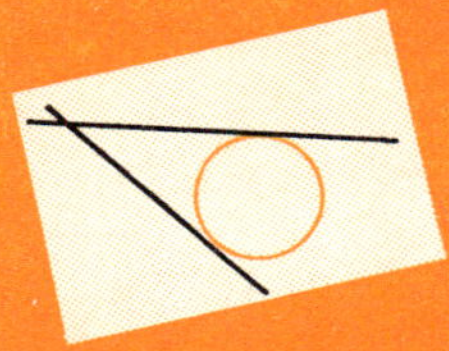

(L^2, C)

The lines on the acetate sheet intersect. The acetate is moved over the circle to produce the main possible configurations.

This situation leads again to various angle properties of the circle. Further properties emerge in the reverse case (C, L^2). See A.T.M. handbook, *Some lessons in mathematics*, pp. 266–70.

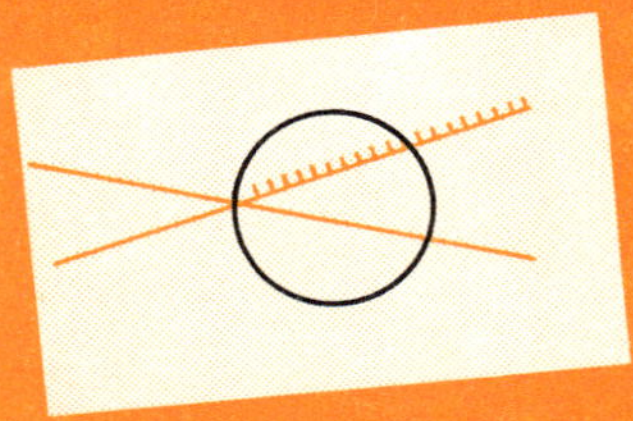

(C, L^2)

Draw the lines on the paper intersecting at about 40° and the circle on the acetate sheet with radius of about 3 cm.

Move the acetate so that the circumference of the circle passes through the point of intersection. Rotate the acetate about this point. Consider the various configurations that arise.

Mark one of the lines at equal intervals, say $\frac{1}{2}$ cm. Repeat the motion, marking those points on the circle which coincide with marks on the line.

Consider relationships between intervals on the line and intervals on the circle (introducing trigonometric functions).

Investigate further situations using points, lines and circles,
e.g. (P, P^2), (P^2, L^2), (C^2, P), (C^2, C^2).

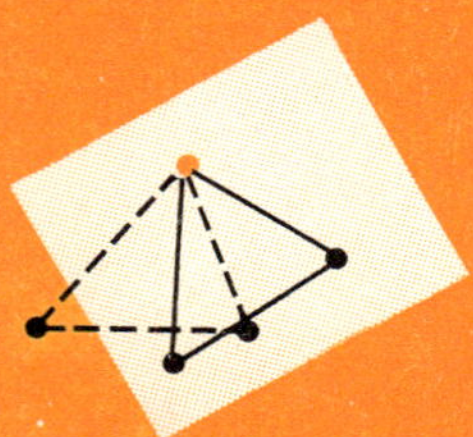
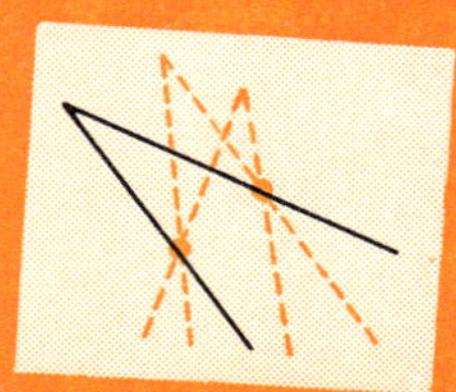

Duplicate some sheets of paper with some different shapes, e.g. triangles, quadrilaterals, regular polygons with, say, 2 cm sides, and so on.

Make a tiling pattern on the acetate sheet using one shape, two shapes and so on.

Which shapes will completely tile (tessellate) the sheet on their own? In pairs? Why?

The patterns on the acetate can be displayed with a white backing sheet.

Duplicate sheets of different tilings—triangles, quadrilaterals, squares, hexagons.

Place the acetate on one of these tilings, and copy one unit shape. Investigate the movements required to map this shape into another position on the sheet.

Draw in any axes of the symmetry the unit shape may have.

What happens to these during the motion? What happens to diagonals?

See p. 126 for further details of possible developments of this situation.

Finally this is a suitable place to remind the reader that when certain grids are printed on acetate sheet and superimposed, a fascinating variety of so-called *moiré* patterns arise. A useful and inexpensive set of prepared sheets can be bought from Proops Bros. Ltd (see p. 233).

arrows

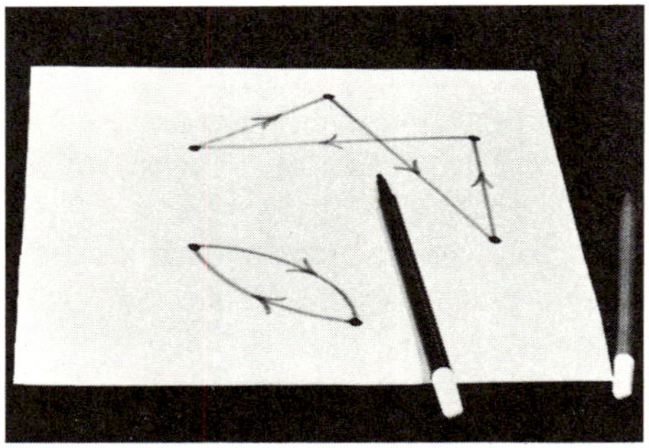

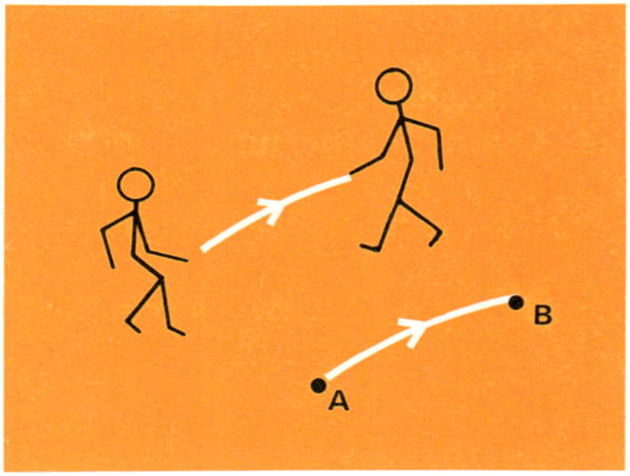

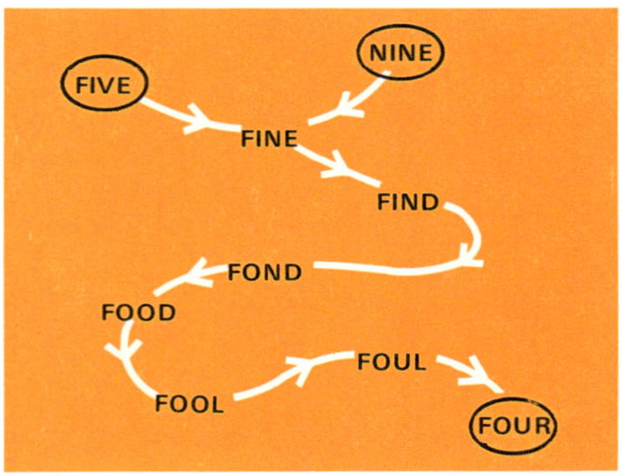

Mathematicians have traditionally recorded their thoughts by making marks on writing surfaces such as sand, stone, paper, blackboards and so on. Here we wish to consider the systematic use of a particular mark—the arrow—which can be very fruitful. Our material is in this case symbolic.

Attractive *arrow-graphs* can be made with felt pens. These pens cost 5p–15p each. Plain paper is clearly most suitable for this work; the cheapest is thin A4 copypaper at about 60p a ream, i.e. 500 sheets.

Apart from the resulting gaiety, the arrows do seem to take on a life of their own. They call attention to the relation between things rather than the things themselves. This makes them useful over a wide range of mathematics. Here we can only indicate briefly some possible developments.

An obvious starting point with young children is to use arrows to symbolise relationships among themselves. For example, the arrow from A to B may mean that A is the brother of B, A has a birthday in the same month as B, A is taller than B, A is not older than B, and so on. The children themselves will invent a rich variety.

A well known example, due to Papy, has an arrow from A to B when A's first name has the same initial as B's surname, e.g. an arrow would go from Sally Brown to John Smith.

Another example makes use of the game of changing the letters of a word one at a time to make another word. Thus FIVE becomes FINE, which can then be changed to NINE. Can you change FIVE to FOUR? What is the least number of ways in which this can be done?
(Notice that arrows can be reversed in the figure.)

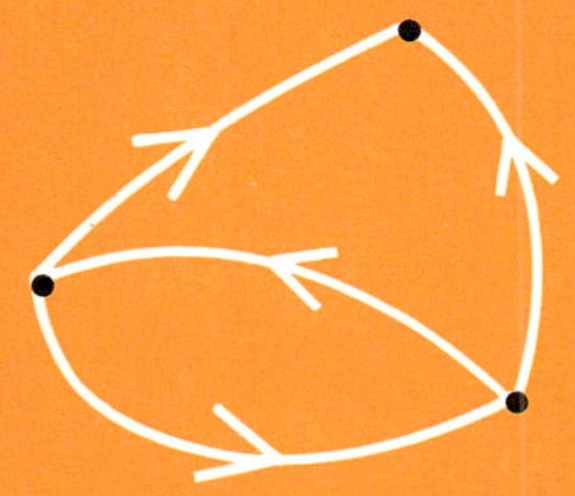

A → B : A is B's brother

The arrow graph for the relation, 'is a brother of', has been drawn for a particular group of four children. What can be deduced from the graph about the children?

A relation may produce a particular pattern in some case but quite a different one in another. Thus for another group of children it turns out that no arrows can be drawn.

What can be deduced about these children?

(An awkwardness of language can often arise, when say, 'Sally is Tom's sister' is forced into forms like 'Tom has as sister Sally'. A possible convention has been indicated. The arrow graph can be titled: A→B means that A is B's sister.)

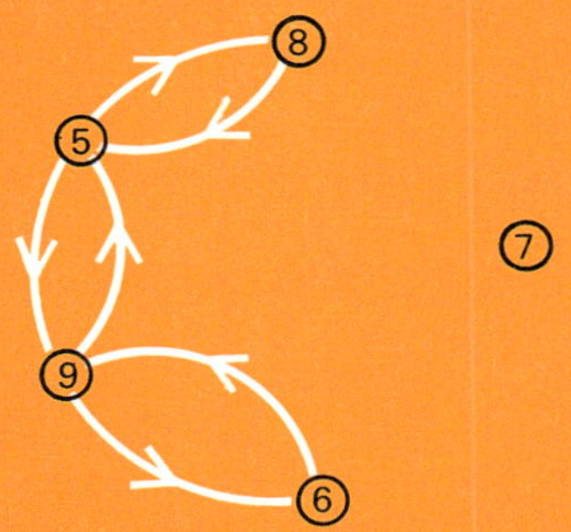

It is natural to consider arrows which link numbers.

The arrow from x to y might mean

y is twice x

x is less than y

y is the square of x

the sum of the first x integers is y.

A useful classroom game is for one person to invent a relation to be guessed by others from a successively extended arrow graph. An example invented by a boy for such a game is: the numerical difference between x and y is not less than 3.

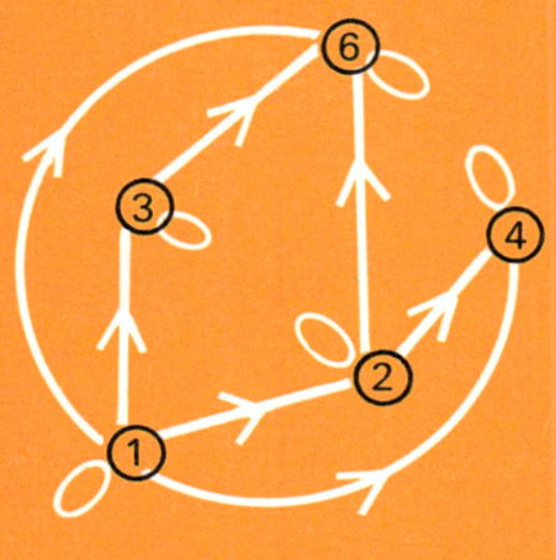

A → B : A is a divisor of B

The use of arrow graphs to prompt further questions is particularly useful in a systematic study of divisibility in a set of integers. Here $m \to n$, m and n are positive integers, might mean

m is a divisor of n

m has n divisors

the sum of the divisors of m other than m itself is n

the number of integers less than and coprime to m is n.

Using an arrow graph as a systematic record of information is often more interesting than making a table. It is worth drawing all the above graphs. Making them will suggest certain conjectures. . . .

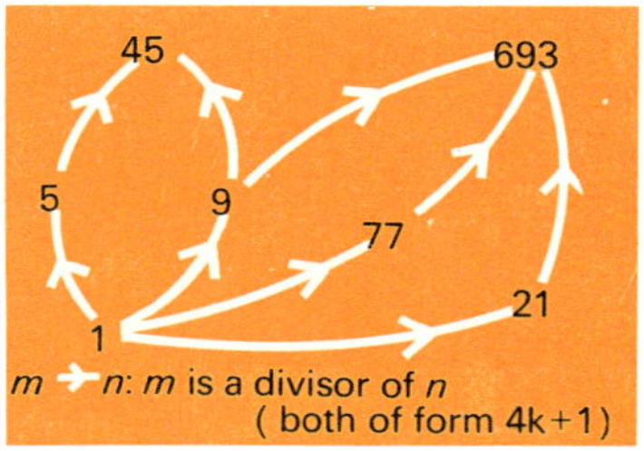

Investigate these graphs for some restricted set of integers. For example, odd numbers only, or numbers of the form $4k+1$, or square numbers.

Divisibility in such restricted sets may be surprising.

What are the *primes* of these sets? Do the numbers factorize uniquely into primes in each case? In the case of numbers of the form $4k+1$ as shown, $693\ (=9\times 77=21\times 33)$ is the product of primes in two different ways.

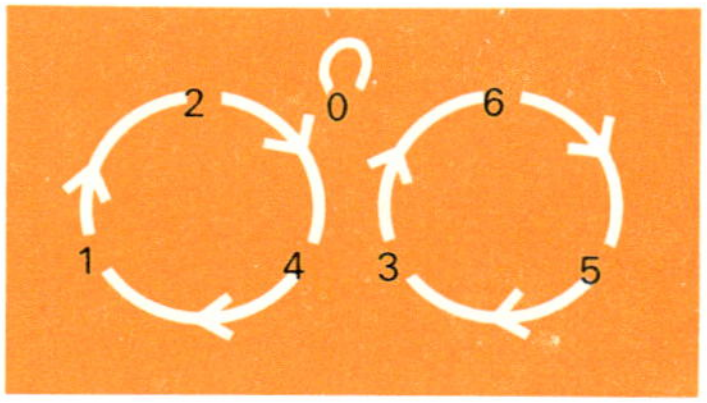

The underlying structure of modular arithmetic is powerfully revealed by drawing graphs. For example, consider arithmetic mod 7, and a graph where $m \rightarrow n$ means n is twice m (mod 7). Apart from the loop on zero this has two cycles.

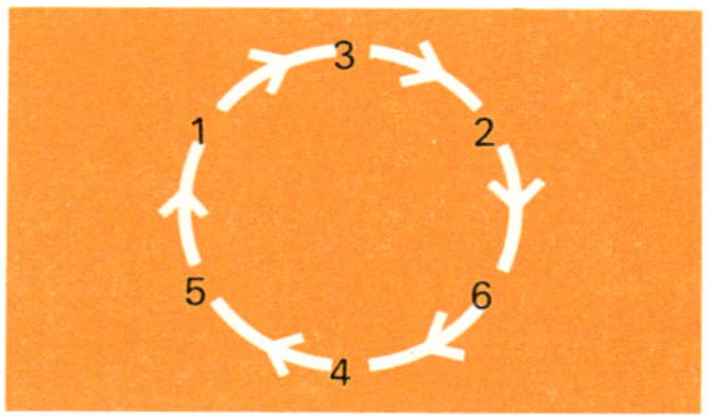

Now draw the graph for multiplication by 3. This has one cycle. Draw the graphs for multiplication by 4, 5, 6. Do they decompose into cycles?

All the graphs can be found within the circular arrangement shown. Can you see how?

Can you organize the multiplication graphs of other arithmetics in this way?

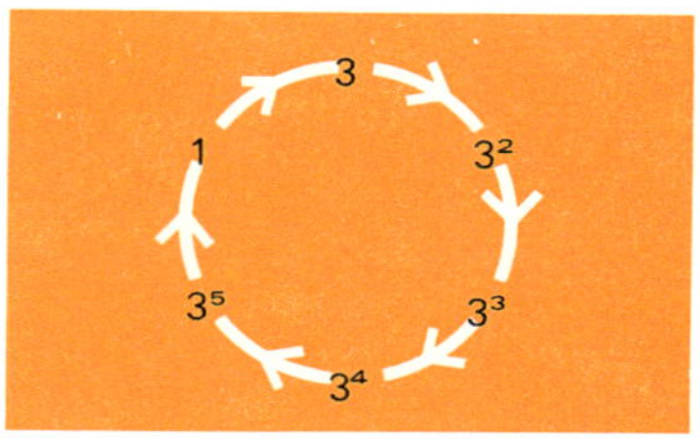

For mod 7, all numbers are powers of 3 and also powers of 5. What are the generators of other arithmetics?

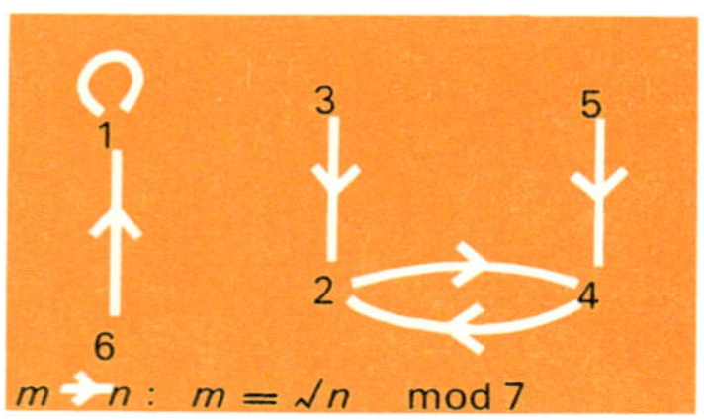

For mod 7, the numbers 1, 2, 4 have square roots, but 3, 5, 6 have not. What about other arithmetics?

For mod 7, the number 6 can be expressed as -1. It has no square root. Is there a $\sqrt{-1}$ in other arithmetics?

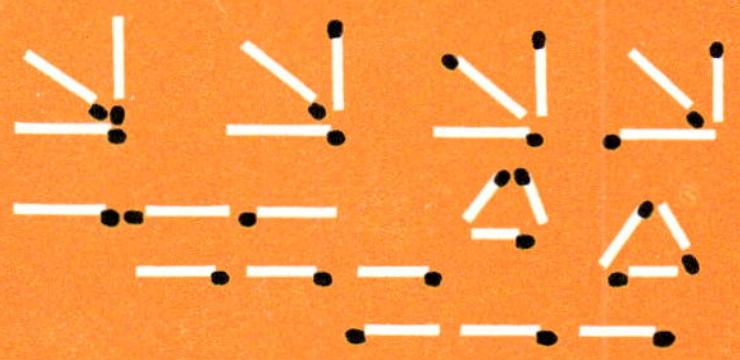

	a	b	c
a	0	1	0
b	1	1	0
c	1	1	0

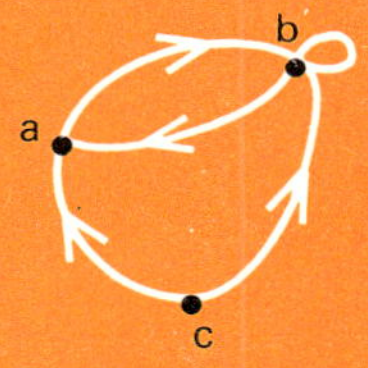

	a	b	c
a	1	1	0
b	1	2	0
c	1	2	0

$$= \begin{pmatrix} 0 & 1 & 0 \\ 1 & 1 & 0 \\ 1 & 1 & 0 \end{pmatrix} \begin{pmatrix} 0 & 1 & 0 \\ 1 & 1 & 0 \\ 1 & 1 & 0 \end{pmatrix}$$

number of routes over two paths

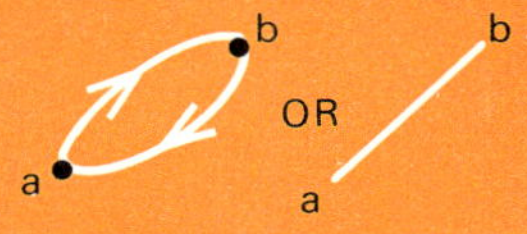

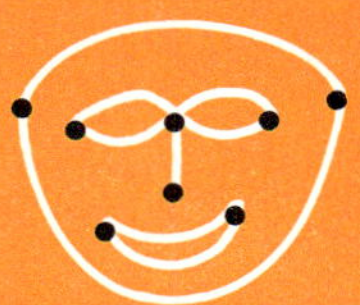

Consider drawing arrow graphs quite abstractly as links between points. How many graphs can you draw with 3 arrows?

How many different graphs?

Take a number of matchsticks. Arrange them on a table with sticks touching end to end. How many different arrangements can be made? The match heads indicate a direction or arrow on each match.

An arrow graph can be represented by a matrix in various ways. One type of *incidence* matrix records the number of arrows from *a* to *b* for each pair of points. This matrix can be combined with itself in a particular way to give the number of paths from *a* to *b* over two arrows. A natural combinatorial argument leads to the rule for so-called multiplication of matrices.

As an example of the use of arrow graphs to illuminate and sometimes solve various problems, see p. 74 or p. 86. Many games and puzzles can be looked at usefully in this way and the reader is reminded of the increasing use of such techniques in solving industrial problems, e.g. critical path analysis.

Where an arrow from *a* to *b* always implies an arrow from *b* to *a* the relation involved is said to be symmetric. In such cases *a* and *b* could be linked without an arrow. (The heads of the matchsticks are chopped off or ignored. There are no one-way streets.) There are plenty of situations with these symmetric graphs to explore. The following samples may indicate the flavour.

Draw a graph. Is it possible to get from any point of the graph to another along lines of the graph? If so the graph is *connected*.

Draw some unconnected graphs, and count the number of points, lines, regions and connected parts of each graph. Can you find a general relationship between the four numbers in each case?

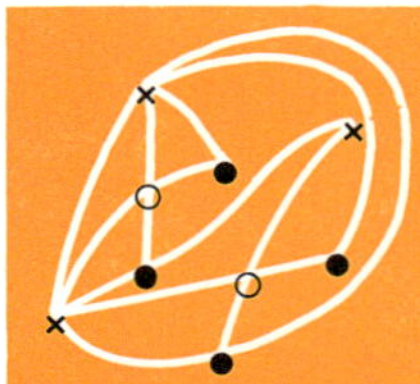

× \ ●	2	3	4	·
2	0	0	0	·
3	0	1	2	·
4	0	2	·	·
·	·	·	·	·

least crossings

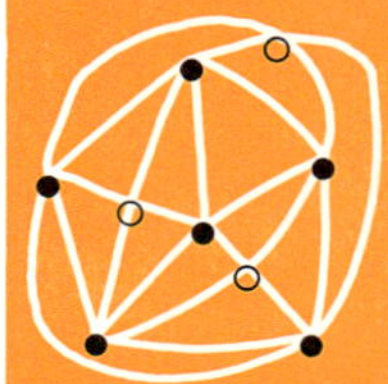

points	least crossing
·	·
4	0
5	1
6	3
·	·

Mark 3 points with small crosses and 4 with small dots. Join dots to crosses in every possible way. What number of intersections do you get? What is the least number you could get? What is the least number for say, 4 crosses and 4 dots? For other numbers of crosses and dots? cf. p. 83.

Mark six points and join them with lines in every possible way. What number of intersections do you get? What is the least number you could get? Try different numbers of initial points. cf. p. 82.

Draw some graphs, and colour the regions so that regions sharing an edge have different colours.

Some graphs need only two colours, some only three. Can you describe what sort of graphs these are in each case?

Do any graphs need four colours?

Five?

Take a number of matchsticks. Arrange them on a table with sticks touching end-to-end. How many different arrangements can be made discounting direction? A useful substitute for matchsticks are wooden lighting spills broken into halves. A packet of 100 spills cost about 10p from tobacconists.

Take four strips and make a *connected* shape with strips meeting only end-to-end. Some shapes will have closed regions in which case there will be a route, along some strips, that is closed.

Consider how many such closed routes the various arrangements have.

Remove and replace strips until there are no closed routes but still a connected arrangement. Such a pattern is called a *tree*. How many different trees can be made with 4 strips? With 5 strips? Can you generalize?

The theory of graphs of this kind—directed and undirected—is a fairly recent, but rapidly growing, subject. It provides plenty of interesting situations for investigations in schools as well as being a useful preliminary to topology.

The secondary textbooks by Papy (see below) develop a mathematics course with a systematic and colourful use of arrow graphs though the uncompromisingly structured course does not seem to leave time for investigation of the actual graphs themselves. Arrow graphs are still only an occasional feature of work in secondary schools. There appears to be a case for a regular use of them at all levels.

some books:

Arnold, B. H., *Intuitive concepts of topology*, Prentice Hall, 1962.

A.T.M., *Some lessons in mathematics*, C.U.P., 1964.

A.T.M., *Notes on maths. in primary schools*, C.U.P., 1967.

Dynkin, E. B. & Uspenski, V. A., *Problems in the theory of numbers*, Heath, 1963.

—, *Map colouring*, Heath, 1963.

Fujii, J. N., *Puzzles and graphs*, N.C.T.M., 1963.

Moon, J. W., *Topics in tournaments*, Holt, 1968.

Ore, O., *Graphs and their uses*, Random House, 1963.

Papy, G., *Modern mathematics*, vols. 1–2, Macmillan, 1968.

books

In this context we are concerned with books that can be considered part of the stock of material kept in a classroom as resources for making mathematics. Single copies of course textbooks may well form part of such a working library. Books that are mainly expository are not the best for the purpose we have in mind, however good they may be in other respects. Nor are we here concerned with the many excellent general background books which should certainly be available in the school library. There is, of course, a wide range of books from which to choose. In particular, we draw the reader's attention to the growing literature of recreational mathematics, though this is variable in quality; also to the short pamphlets for children on a single mathematical theme. There are more of these being published now, though nothing like enough, and few of them yet succeed in catching the right note.

Lists of books can be very daunting and we only attempt a brief selection from limited experience. We give a list of about forty items, mainly concentrating on books or pamphlets that we think can be usefully consulted by pupils directly.

There are various further sources of information worth mentioning. Various journals publish regular book reviews; the Mathematical Association has a useful school library list; catalogues of books on mathematics currently available are regularly prepared and distributed without charge by major booksellers (some addresses on p. 232). The Institute of Education library of Southampton University has prepared a useful annotated bibliography of books on modern mathematics (1966, 25p including postage). We remind the reader that the library facilities of the local Institute of Education are available to teachers in the area.

Ball, W. W. R., *Mathematical recreations*, Macmillan, 1959.
Baravelle, H. v., *Geometry*, Waldorf School, N.Y., 1948.
Barr, S., *A miscellany of puzzles*, Crowell, N.Y., 1965.
Barr, S., *Experiments in topology*, Crowell, N.Y., 1964.
Beiler, A. H., *Recreations in number theory*, Dover, 1963.
Boeke, K., *Cosmic view*, Faber, 1949.
Cole, J. P. & Beynon, N. J., *New ways in geography*, Blackwell, 1968.
Cundy, H. M. & Rollett, A. P., *Mathematical models*, O.U.P., 1961.
Davis, P. J., *Lore of large numbers*, Random House, 1967.
Domoryad, A. P., *Mathematical games*, Pergamon, 1964.
Ehrenfeucht, A., *The cube made interesting*, Pergamon, 1964.
Escher, M. C., *Graphic work*, Oldbourne, 1961.
Fielker, D., *Topics for maths.*—various titles, C.U.P., 1969–.
Fletcher, T. J., *0 and 1*, Mathematical Pie, 1961.
French, P., *Exploring maths.*—various titles, McGraw Hill, 1963–.
Gardner, M., *Maths. puzzles and diversions*, Penguin, 1965.
Gardner, M., *More maths. puzzles and diversions*, Penguin, 1966.
Giles, G., *Switch boards and logical paths*, A.T.M., 1970.
Golomb, S., *Polyominoes*, Allen & Unwin, 1966.
Graham, L. A., *Surprise attack in maths.*, Dover, 1968.
Johnson, D. A. et al., *Exploring maths.*—18 titles, Murray, 1961–.
Juster, N., *The dot and the line*, Nelson, 1963.
Land, F., *Language of mathematics*, Murray, 1961.
Lewis, K., *Experiments in maths.*—7 titles, Longmans, 1966–.
Lietzmann, W., *Visual topology*, Chatto and Windus, 1965.
Lockwood, E. H., *A book of curves*, C.U.P., 1967.
Midland Maths. Experiment, *Excursions*, 2 vols., Harrap, 1968.
Mold, J., *Topics for maths.*—various titles, C.U.P., 1969–.
Mott-Smith, G., *Mathematical puzzles*, Dover, 1968.
Ore, O., *Graphs and their uses*, Random House, 1967.
Paling, D. et al., *Making maths.*—12 topic books, O.U.P., 1969–.
Peck, L. C., *Secret codes*, etc., N.C.T.M., 1961.
Polya, G., *Mathematical discovery*, 2 vols., Wiley, 1964.
Rowland, K., *Looking and seeing*, 4 vols., Ginn, 1964–.
Sackson, S., *A gamut of games*, Random House, 1969.
Steinhaus, H., *Mathematical snapshots*, O.U.P., 1960.
Steinhaus, H., *100 problems in elementary maths.*, Pergamon, 1963.
Stover, D. W., *Mosaics*, Houghton-Mifflin, 1966.
Tahta, D. G., *Pegboard games*, A.T.M., 1967.
Thompson, D. W., *On growth and form*, C.U.P., 1961.
White, G., *A world of pattern*, Murray, 1961.
Zippin, L., *Uses of infinity*, Random House, 1965.

calculating machines

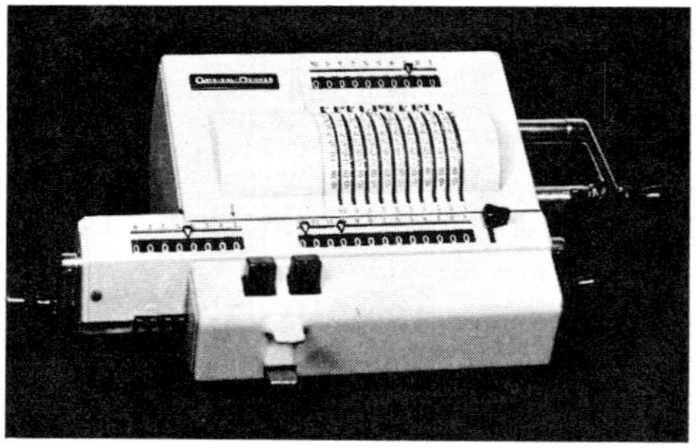

There are a considerable number of desk (hand) calculators now available. A full account of some machines and their capacities can be found in a supplement to the A.T.M. journal, no. 31. A basic lever-set or key-set machine can be purchased for about £50. Most of them are imported and one of the important things to look for is the service facility available.

Although the key-set machine is quicker, and so more useful for, say, statistical work later in the secondary school, the lever-set machine offers more possibilities for the whole ability range. The levers emphasize place-value, and their movement gives a feel for the size of the numbers used. At the early stage, children should be given freedom to explore the machine. With very little guidance, the basic operations are soon mastered even by very young children. We have often found it useful to ask for the number of operations involved in any process that has been tackled on the machine. There is the implied possibility that it might be done in less, and this not only stimulates further exploration of the machine, but also of other arithmetic processes.

It is not necessary to have many of these machines. If each child can become familiar with one during his first year in the secondary school, then it will be used throughout the course, when and as required. This accepts the principle of *not* keeping the machine under lock and key.

We have given an account of a possible investigation that might arise with a calculating machine on p. 84. Regular use of a machine may involve the making of flow charts and some simple programming. It will also draw attention to iterative methods for such processes as obtaining a square root or solving a quadratic equation.

Calculating machines clearly lead on to computers. We do not think it appropriate to include computers in our list of materials, for these will certainly involve a special budget and special facilities for some time yet. But there are various ancillary items that can be used quite simply. Punched cards can be made with ordinary postcards and a filing punch; if some of the holes along one edge are cut open, a pack of

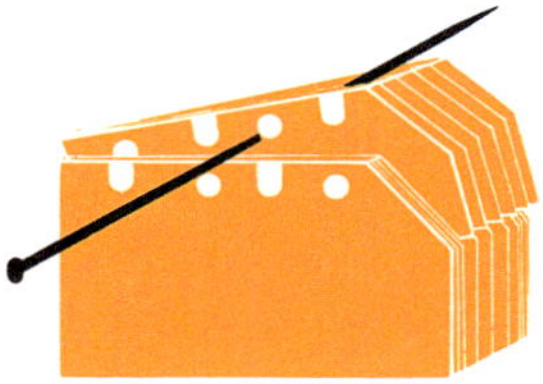

cards can be sorted by inserting knitting needles, or compass points, at appropriate places. Punched paper tape can be obtained from any computer installation and presents a fascinating decoding exercise as such. Another simple logical device is the circuit board, briefly discussed on p. 178.

The National Computing Centre (for address, see p. 232) produce a computer appreciation 'package' for schools which includes notes, transparencies and game material.

Closer links with the reality of computing may be possible if there is a computer installation in the area. In particular, the interested teacher should inquire at the local technical college or university where computer time may be readily available. Some schools already enjoy a computing service, at little cost, by post. It is clear that computers will be ordinary working tools for many pupils who already take to them quite naturally when they are available. They may well change school mathematics quite drastically. Exactly what place they will occupy in a mathematics course will depend on the teacher's total conception of what his pupils are studying mathematics for. But the question is likely to be one of the most important, and possibly controversial, issues in mathematical education in the next few decades.

some books:

A.T.M., *Some lessons in mathematics*, C.U.P., 1964.
Bolt, A. B. & Wardle, M. E., *Communicating with a computer*, C.U.P., 1970.
Bowden, B. V., *Faster than thought*, Pitman, 1953.
Broderick, W. R., *The computer in school*, Bodley Head, 1968.
Fielker, D., *Computers*—topic pamphlet for pupils, C.U.P., 1968.
Hamming, R. W., *Calculus & the computer revolution*, Houghton Mifflin, 1968.
Knuth, D. E., *Fundamental algorithms*, Addison Wesley, 1968.
Lewis, W. D., *Teaching school mathematics with a desk calculator*, Heinemann, 1966.
Moakes, A. J., *Numerical mathematics*, Macmillan, 1963.

circles

There are many ways in which circles can be studied. It is useful, in the first instance, to be able to make many different circles quickly and easily. Ordinary compasses tend to loosen and children have difficulty in finding the right pressure. An alternative is the draughtsman's beam compass, which has the advantage of drawing attention to the constant radius property. Inexpensive plastic versions are available from E.S.A. (for address, see p. 232).

Circular counters made of cardboard or plastic are available in different sizes. A quantity of these would be useful for actual manipulation and play.

Coins make useful units—the 1p piece is approximately 2 cm diameter. Drawing round counters or coins is often a quick way of drawing circles, satisfactory for many purposes. It is easier to draw round the inside of a stencil and it could be useful to have a set of card or plastic stencils with circular holes of various sizes.

Coloured gummed paper circles are attractive and cheap. Packets of 500 in various sizes from 2 cm to 30 cm diameter cost from 10p to 50p a packet from Charles Ltd (see p. 232). This firm also produces various regular polygons.

Circular geoboards, the Spirograph and various films, provide some of the many other ways in which circles can be systematically investigated. We have given one starting point, using a film, on p. 150 and add some brief suggestions here.

Free play should be encouraged with all these materials. Take some circles. . . .

Various rules or constraints on the making of circles are possible and the consequences of such rules can be explored.

Make some patterns of circles starting from a given circle, a given square, and so on.

Make some circles whose centres lie on some given locus. Consider these circles that satisfy a further condition, e.g. pass through a point.

Thus the circles, whose centres lie on a fixed given circle and which pass through a point on this circle, all touch a cardioid.

Make up and explore other conditions.

Make some circles that pass through two points—using counters, coins or gummed paper shapes.

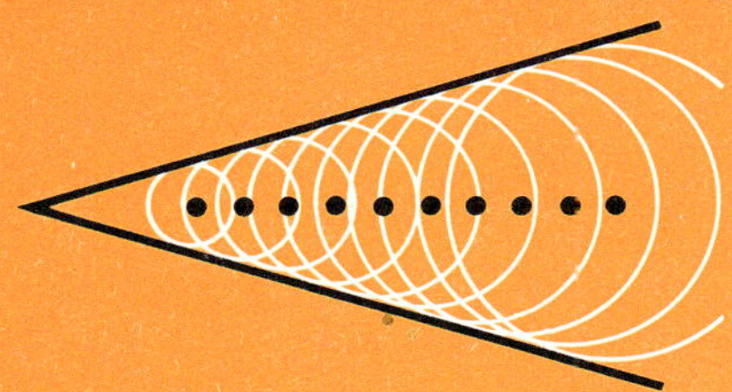

Make some circles that touch two fixed lines.

Make some circles that touch two fixed circles, or a line and a circle, and so on.

Where do the centres of these circles lie?

These situations make attractive subjects for a short film.

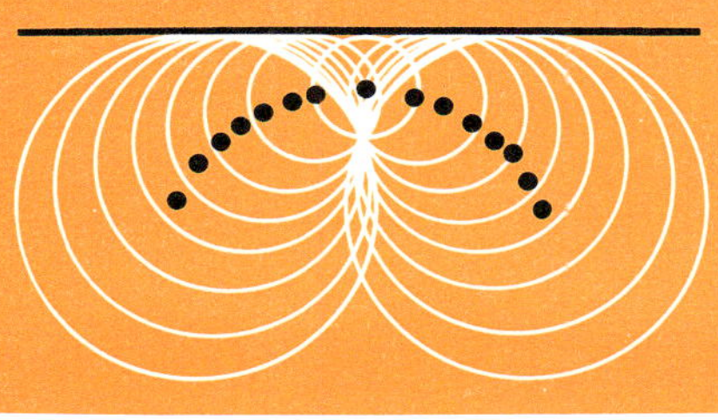

The centres of the circles touching a fixed line and passing through a fixed point define a parabola. Various properties of the parabola are immediately available from this definition. This and the generation of other conics in this way are shown in some of the short 16 mm films by J. L. Nicolet, distributed by Cuisenaire (see p. 232).

How many circles can be drawn satisfying three conditions, e.g. passing through 3 points, touching 1 circle and 2 lines, touching 3 circles and so on? How can the circles be constructed in each case? (This is known as Appollonius' problem—there are accounts in various books, e.g. Coxeter, *Introduction to Geometry*, Wiley, 1961).

Take some counters, coins or paper circles of the same size, and see how they pack together without overlapping. What regular tilings are possible?

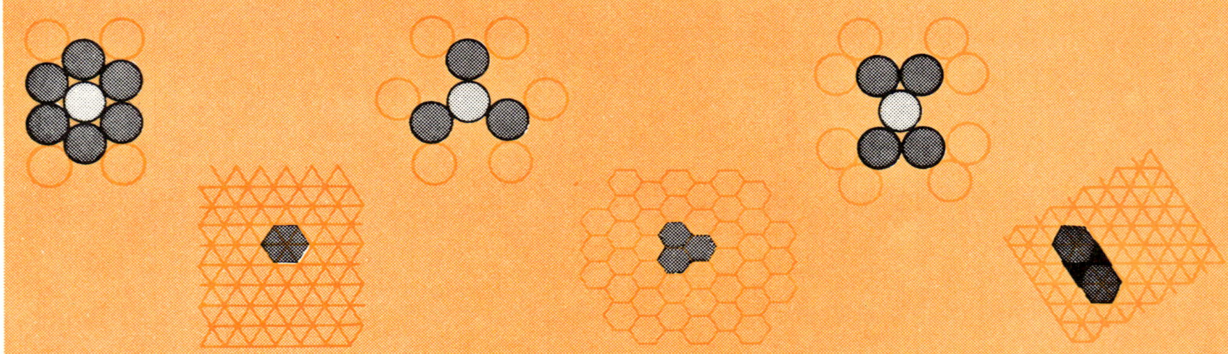

In the packing of tins of food or the manufacture of cables of wire, it is useful to know the smallest area in which a certain number of circles will fit.

Take some discs and find the smallest circle or the smallest square in which they can be packed.

Make a collection of possible arrangements.

Investigate the regular patterns that can be made by packing two sizes of disc.

What regular patterns can be made with overlapping circles?

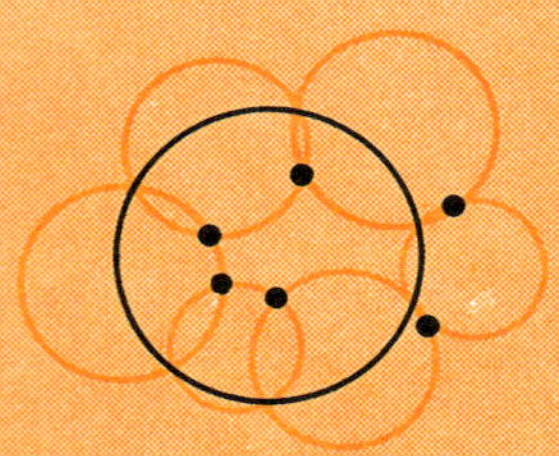

Draw a circle and make a closed chain of circles each intersecting its two neighbours such that one of the points of intersection of each pair of neighbours lies on the first circle.

How do the other points of intersection lie?

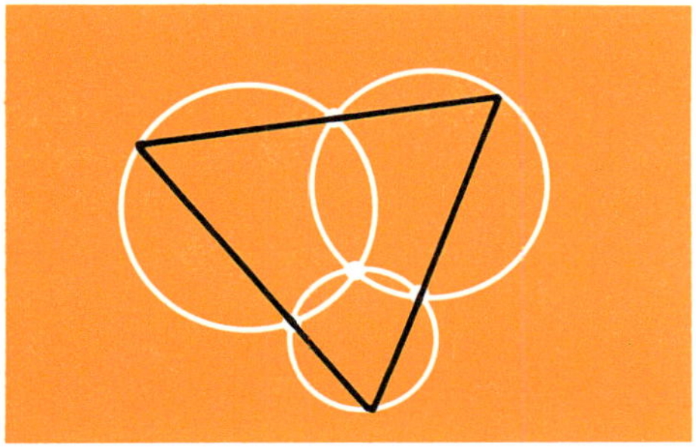

Draw three circles through a point. Can you make a triangle—with a vertex on each circle and sides passing through the other points of intersection of the circles? Investigate the corresponding situation for various numbers and sizes of circles through a point.

How many discs of the same size can be placed each covering the centre of any one of the others?

(Any number of discs could be stacked in a pile to satisfy the conditions but what if no discs can be directly over another?)

Investigate other covering problems.

Investigate the area of circular discs by making circles on square grids. Given a unit disc what is the area of one that has twice its diameter? Three times its diameter?

The important point here is that it may be more useful to think of the familiar area formula as $r^2\pi$ where π is the area of the unit disc.

Approximations to π may be found by counting squares of the grid or by polygonal approximations.

Consider expressing all areas in terms of circular rather than square units. Thus the area of a circle radius r is r^2 circular units. What is the area of a square, a triangle in these units?

Circles could lead on to spheres. Polystyrene balls in various diameters may be obtained from Elford Plastics Ltd (see p. 232).

circuit boards

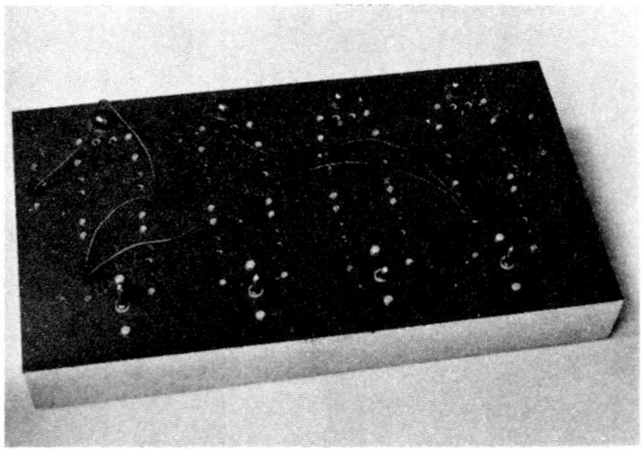

A circuit board provides an interesting introduction to logic and Boolean algebra. Home-made versions can be made by pupils—or teachers—who can read a simple wiring diagram, and there are various designs with full instructions available (see the A.T.M. books listed below). The electrical parts required can be bought from wholesale firms such as Radiospares for about £3. The board may be powered by mains supply through a transformer, or by battery. The latter makes the unit more mobile; some spare batteries and bulbs will save considerable frustration.

A ready-made board can be purchased from Tor Instruments (see p. 232) for about £5·50.

At first the board should be available for unhurried exploration, even to the extent of letting the pupil find out for himself which wiring arrangements produce a light. The layout demands certain questions of the investigator, e.g. how can two lights be made to come on with one switch?, how can one light be made to come on with two switches?, and so on.

Extensive selections of problems to be explored on the circuit board are given in the A.T.M. handbook and pamphlet listed below.

E.g. A light is to be operated from 2 switches, 3 switches, . . . Wire up accordingly.

Wire up the board so that when a number between 0 and 15 is set up in binary notation with 4 switches, a light comes on if and only if the number is prime.

The board clearly provides an interesting model for set notation and various logical operations. Alternatively, these can be derived from the wiring. In tackling problems, children often work directly with the patch wires, rather than translating from some other way of symbolizing the logic of the situation. Two basic configurations of wires will appear over and over again. These correspond to the logical operations AND and OR ELSE. But they may become the symbols of the logic directly.

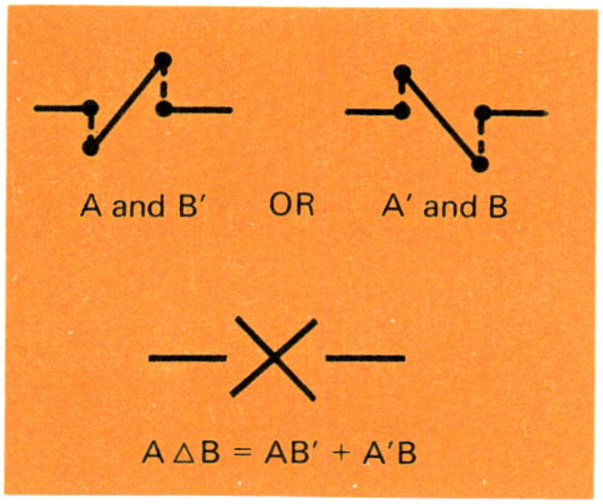

Thus consider the example of a light to be controlled by two switches, say A and B. We might argue that the light will have to come on if and only if the switches are in certain states. Alternatively the argument may develop entirely in terms of wires. Some symbolic representations of various ways of thinking are shown.

It is often useful to ask for a problem to be set up with least number of wires. This minimizing is an important feature of circuit design and implies various identities of Boolean algebra. Thus in the case of the light controlled by three switches, A, B and C, the appropriate Boolean expression is $AB'C+A'BC'+A'B'C+ABC$, and there are various ways of factorizing this. Working directly from the wires may have yielded a wiring more appropriately symbolized as $A \triangle B \triangle C$, where $\triangle$ is the (associative) symmetric difference operation.

A NAND/NOR logic board provides an interesting challenge. A useful board is produced by Tor Instruments for £17·50. The logic here may seem unfamiliar to the adult, yet the child working directly on the board will soon become operational. In this case all the lights are on when the power is switched on and an input to a light will change its state. An immediate challenge seems to be to wire the board so that all the lights are off, or that some designated ones are off. When the lights are joined in a cycle there should theoretically be alternate switching of states, but in practice there is a dominance pattern, and it is an interesting problem to explore this.

We remind the reader that there are various puzzles involving logic that do not need electrical equipment. For example: make 3 black and 2 white hats. Place 3 children in file and put a hat on each from behind. With white hats on the front two, can the third determine the colour of his hat? If he does so correctly can the middle child do the same? The front child? Vary the combination of colours.

some books:

A.T.M., *Some lessons in mathematics*, C.U.P., 1964.
A.T.M., *Teaching aids and logic*, A.T.M., 1963.
Gardner, M., *Logic machines*, McGraw Hill, 1958.
Giles, G., *Switch boards and logical paths*, A.T.M., 1970.
Wilkinson, A., *Computing models*, Arnold, 1969.

cubes

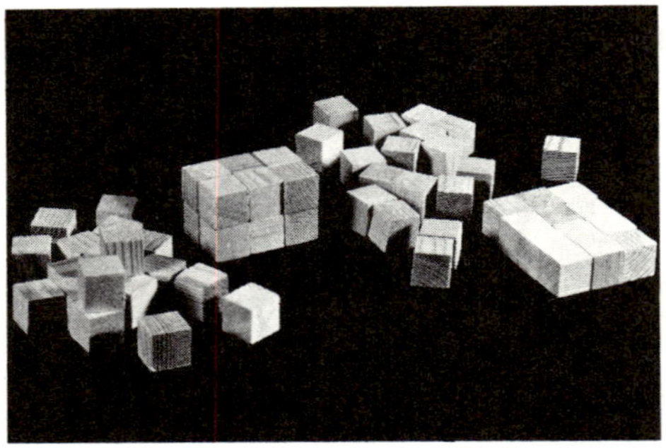

A good supply of wooden cubes is essential. The cheapest way of getting a stock of about a thousand is to order say ten 2 metre lengths of 2 cm × 2 cm soft wood (e.g. red deal) from a timber merchant. It is important to order 2 cm × 2 cm *finished*. The lengths can now be cut with a circular saw into 2 cm cubes (by the woodwork department?)

A set of a thousand cubes are obtainable from Burdett (see p. 232) for about £5. Packets of a hundred 1 cm cubes are obtainable from Brown Ltd (see p. 232) for about 20p.

A useful adhesive for sticking cubes together is Evostick.

When required the faces can be marked with felt pen, pencil, or chalk.

Take a number of cubes. Can you assemble these cubes to form a single cube? If not, how many more would you need in order to make a cube?

What numbers of cubes can be put together to make larger cubes?

Make a 6 × 4 × 2 cuboid. Divide the cuboid into layers. How many cubes in each layer? How many layers? Do this in different ways.

Using cartridge paper or thin card, make a rectangular box with a volume of 36 cubic units. Check the volume by filling it with 36 unit cubes.

Make several cubes with a volume of 36 cubic units. Investigate the relationship between the dimensions and the surface areas of these cuboids. When is the surface area a minimum?

Mark 6 squares on a piece of squared paper as shown. Place a cube on each square to form a cuboid. What is the volume of this cuboid? Place a cube on top of each cube. What is the volume of the resulting cuboid? Repeat.

Investigate the volumes of solids formed by building up layers in this way, from different bases.

How many different shapes can be made using 3 cubes? 4 cubes?

Some ways of combining cubes are shown in the figure. Are there any other different shapes that can be made using 3 or 4 cubes? Make the shapes shown by sticking cubes together. Can you assemble these pieces to make a cube?

This is sometimes known as the *Soma cube* puzzle. A similar puzzle is currently being sold under the name *Cube-it*.

Explore the symmetry of a cube by placing a labelled cube into a 'box' in all possible ways.

Place cubes made out of 1, 8, 27 cubes on a square lattice as shown. Remove all but the bottom layer of each cube and distribute the layers moved in the square grid. It turns out that: $1^3+2^3+3^3=(1+2+3)^2$

Does this result generalize?

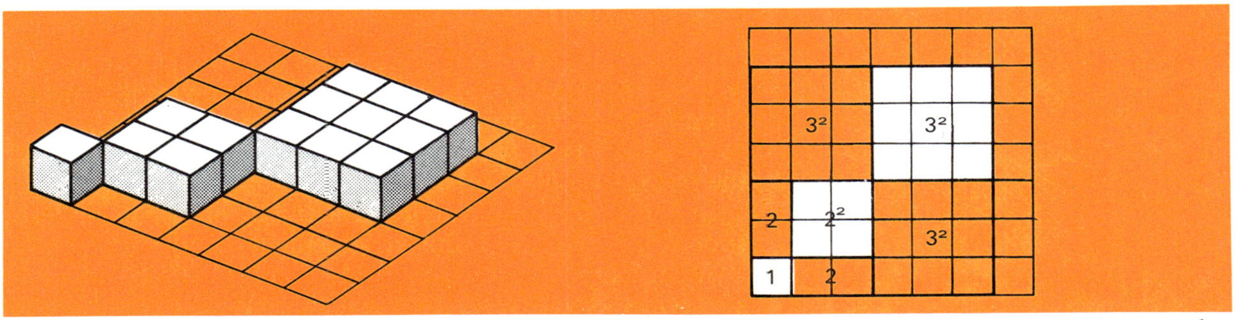

Can you find two cubes that will reassemble to make another cube? (A calculating machine would be useful here. But no cube is the sum of two cubes—a particular case of Fermat's famous last theorem.)

Form pyramids of cubes by stacking layers of triangular numbers of cubes or square numbers of cubes. The number of cubes in each successive pyramid generates a sequence of numbers. Investigate these sequences.

	1	1	1	1
	1	2	3	4
T	1	3	6	10
(P)	1	4	10	20

	2	2	2	2
	1	3	5	7
S	1	4	9	16
	1	5	14	30

Some interesting patterns emerge from these sequences. Thus consider the triangular numbers (T) and the derived pyramidal numbers (P).

T: 1 3 6 10 . . .
P: 1 4 10 20 . . .

Each number in P is the sum of the corresponding number in T and the preceeding number in P. The reader may recognize the pattern of Pascal numbers.

The square numbers S and the derived pyramidal numbers produce an interesting extension of the Pascal pattern.

Investigate further.

Arrange 14 ($=1^2+2^2+3^2$) cubes as shown. The arrangement is indicated by the triangle $a_1b_1c_1$ showing the number of cubes in each column.

Make a further two arrangements orientated as shown by the triangles $a_2b_2c_2$ and $a_3b_3c_3$.

Superimpose the arrangement $a_2b_2c_2$ on $a_1b_1c_1$ and then superimpose $a_3b_3c_3$ on this arrangement.

What happens?

$3(1^2+2^2+3^2)$ cubes give $(1+2+3)$ columns each containing $(2\times 3+1)$ cubes.

A similar arrangement for $1^2+2^2+3^2+4^2$ gives $(1+2+3+4)$ columns each containing $(2\times 4+1)$ cubes.

Generalize for the sum of the first n squares.

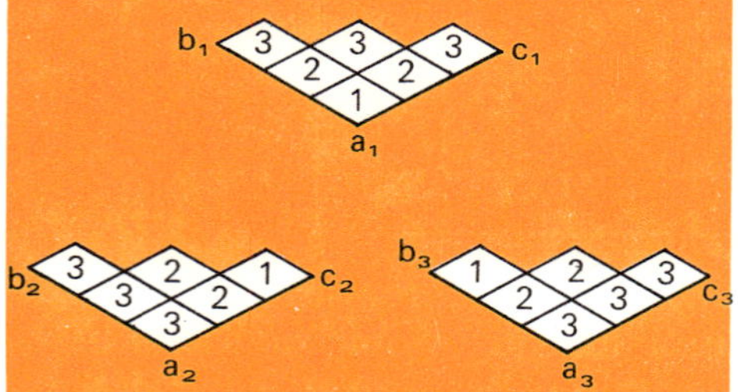

What is the standard way of numbering a die? How many different ways are there of numbering a die if numbers on opposite faces must have a sum of 7?

Mark the faces of a cube with numbers 1 to 6 and place it on a grid of squares the same size as a face of the cube. Tilt the cube successively about edges lying on the grid and label the square of the grid with the number of the uppermost face when the cube lies on the square.

What numberings of the grid are possible?

Investigate for different ways of numbering the cube.

In the same way, tilt the cube successively about the edges lying on the grid until the 6 is uppermost again. Label the square which the cube is now on with the number of tilts made. Repeat for other squares of the grid, and investigate the resulting pattern of numbers.

What is the minimum number of colours needed to colour the faces of the cube so that no two adjacent faces are the same colour?

How many different ways of colouring a cube are there using 4 colours? 5 colours?

There are 30 ways of colouring the 6 faces of a cube using each of 6 different colours once. Make a collection of the 30 different cubes.

Take one of them and try to arrange 8 others into a cube coloured in the same way as the one chosen and with similarly coloured faces in contact.

How many essentially different cubes are there with
4 red and 2 black faces?
3 red and 3 black faces?

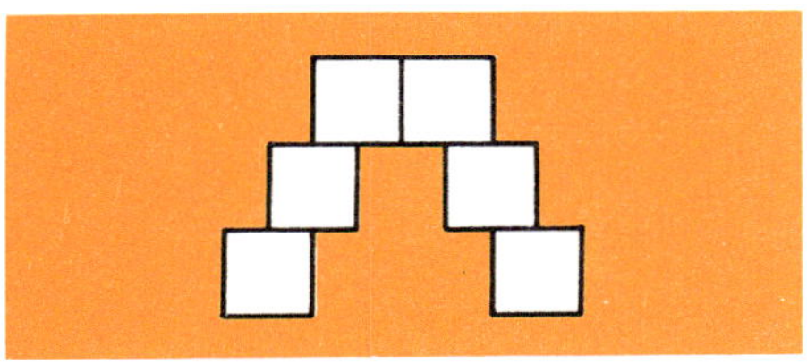

Make a symmetrical arch with some cubes (the cubes must be balanced not stuck). What is the widest arch which can be made with a given number of cubes?

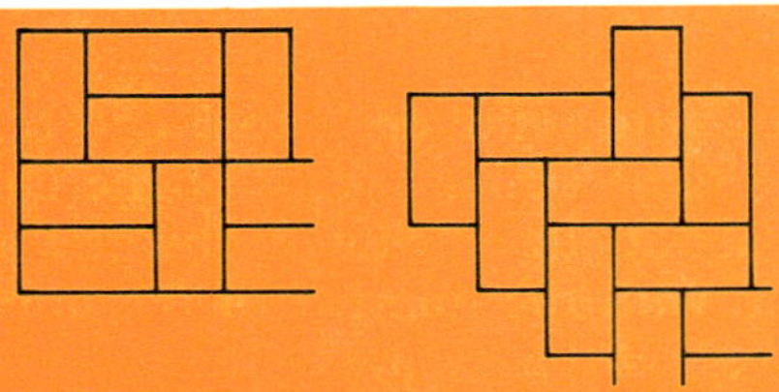

Make some 2 × 1 × 1 units by sticking cubes together. Use these units to make a wall. What bonding patterns can be found? (See Golomb's book listed on page 171.)

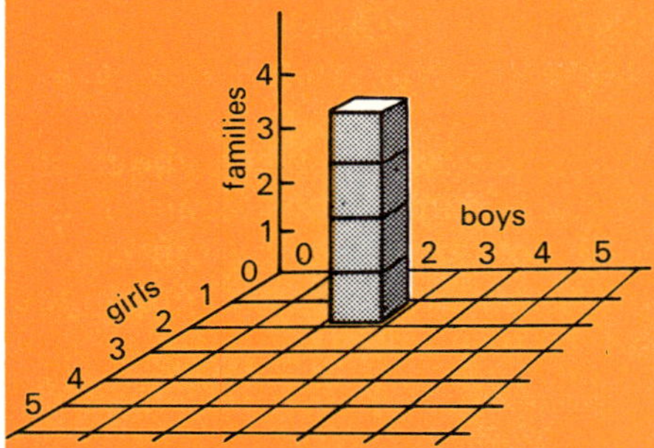

Cubes are useful for building 3-dimensional graphs. For example, a 3-dimensional graph showing the numbers of families with various combinations of boys and girls can be made by drawing axes in a plane to represent the number of boys and the number of girls in each family, and using columns of cubes to represent the number of families with given combinations. (Cf. p. 85). The diagram shows 4 families with 2 boys and 1 girl.

Can a 6 × 6 × 6 cube be made from 27 pieces 1 × 2 × 4?

Take a cube. Mark at least one of its faces—you may mark all the faces if you wish. Repeat with another cube in the same or different manner. Can you place the two cubes together so that each of the four rectangular sides have one and only one marked face? Which initial ways of marking make this possible?

Take four cubes and find ways of colouring their faces using up to 4 different colours in such a way that when placed together in a row each of the four rectangular sides—top, bottom, and two sides of the row—show all 4 colours. Pick up and shake the cubes. Can you reassemble the pattern?

A puzzle of this type is currently being sold under the name *Instant Insanity*.

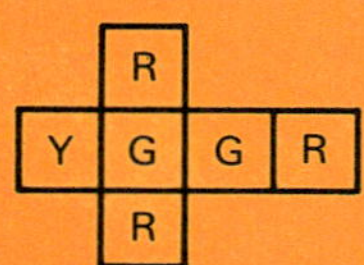

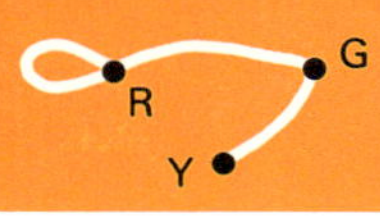

The colouring of a cube can be represented graphically. The colours used are represented by points, and a line is drawn joining two points to indicate the colouring of opposite faces. A colouring is represented by a figure with three such lines. Can this be used to help solve the previous problems?

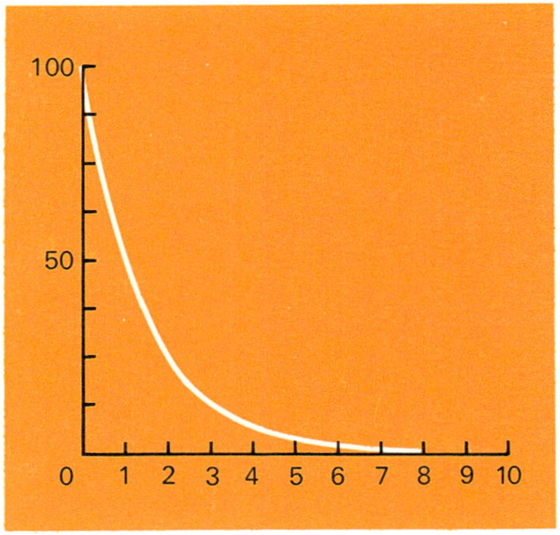

Take 100 cubes. Mark any two faces on each. Place the cubes in a suitable container, shake well and tip out. Remove those cubes which lie with a marked face uppermost. Count and replace the other cubes in the container and then repeat the process. Continue to do this and draw a graph showing the number of cubes remaining in the container at each successive stage. This will approximate to an exponential graph.

What is the chance of a cube falling with the marked face uppermost?

How many cubes should remain after the second shake and tip? The third? The n-th?

How many repetitions of the process are required to empty the container?

What if only one face of each cube had been marked?

Consider each of the cases where 3, 4, 5 or 6 faces are marked.

some books:

Ehrenfeucht, A., *The cube made interesting*, Pergamon, 1964.
Fielker, D., *Topics for Mathematics: Cubes*, C.U.P., 1970.
Gardner, M., *More mathematical puzzles and diversions*, Penguin, 1966.

cuisenaire rods

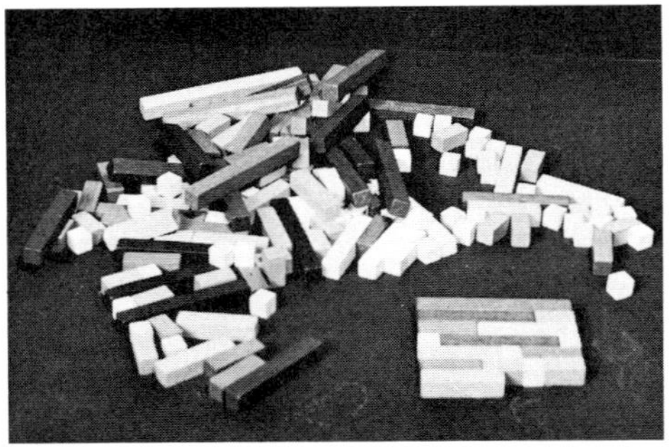

This well known material is not yet widely used in secondary schools, but any study of the available literature on the ways in which it has been employed by teachers and pupils, leaves no doubt that it is as relevant to the sixth form as it is to the infant.

The way in which the material is used as a model for the integers emphasises an algebraic point of view from the start. The exploration of the relations between various rods provides concrete models for a satisfactory development of negative and rational numbers. Using the rods to explore various combinatorial problems, young children have been able to develop parts of mathematics traditionally reserved for a later stage.

A box of rods, suitable in most cases for a group of about four pupils costs £1·35 from Cuisenaire (see p. 232). The magnificent set of squares, cubes and other blocks, exploiting the same colour coding as the rods costs £21·50.

We give a few sample activities to indicate some of the work that can arise. For further details, we refer the reader to the books listed below.

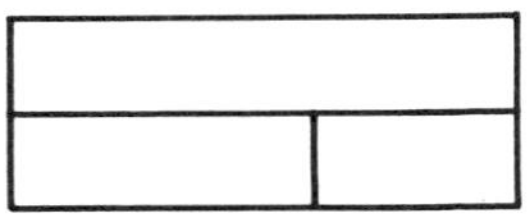

The three rods shown form a basic pattern that can be looked at in many ways.

What rod is equivalent to two others placed end to end?

What pairs of rods can be placed end to end to be equivalent to a given rod?

What rod will complete the gap implied by two others?

Coding the rods by the initial letters of their colours, these ways of looking at the basic pattern may be symbolized in various ways. Where the joining of rods end to end is denoted (by convention) by *plus* we have, for example, $r+g=y$, $y-g=r$, $g=y-r$ and so on.

Choose pairs of rods.

Are any pairs the same? In what sense are they the same? How would you show this?

There are many ways of choosing to call two pairs of rods the same. In particular, certain choices could lead to an exploration of the negative and rational numbers as ordered pairs of natural numbers.

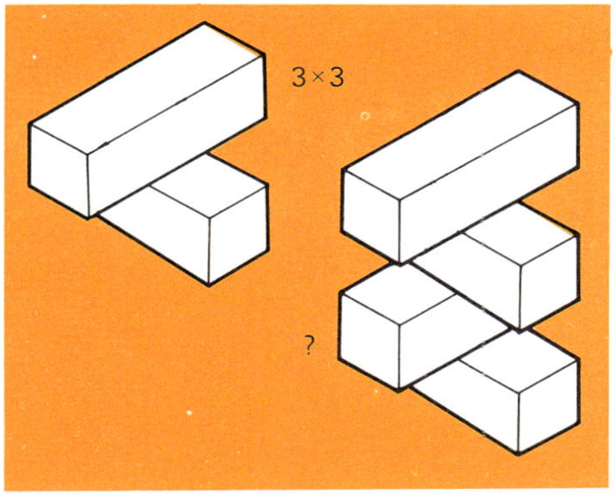

When the smallest (white) rod is chosen as a unit, multiplication of integers is represented, by convention, by laying one rod on top of another.

Four green rods are piled on top of each other. What does the tower represent?

A useful model for indices and logarithms is available.

A line of rods placed end to end is called a train. Make some trains equivalent to a chosen rod, i.e. of the same length. How many different trains can you make? Suppose arrangements of rods in different orders are counted as forming different trains. How many different trains can you make equivalent to each rod?

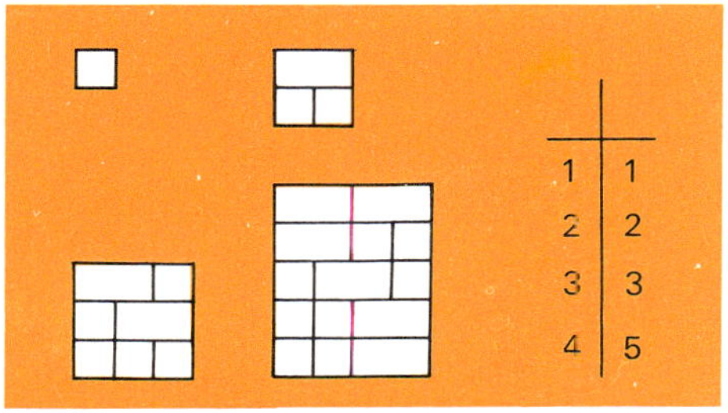

How many different trains can you make using only white and red rods (these are the two smallest sizes, the white is 1 cm long, the red 2 cm long)?

Make a general conjecture from any pattern that you can observe. Can you prove your conjecture?

What difference does it make when different orders are counted as being the same train?

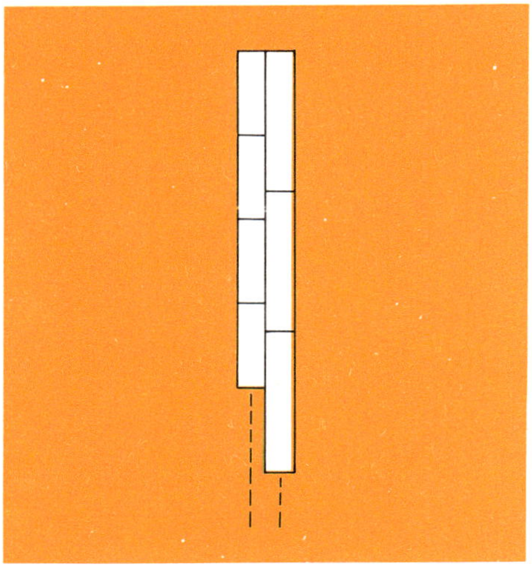

Make various trains of rods using only two colours.

What lengths can be made from different pairs of colours?

Make a train of rods of the same colour and another train of a different colour. Can a train in one colour be the same length as a train of another?

Consider the differences between pairs of trains. What lengths can be expressed as differences of trains in two colours?

Investigate these situations with different pairs of colours.

some books:

Gattegno, C., *Mathematics with numbers in colour*, Bks 1–6, Educational Explorers, 1964, from 25p to 35p each. Conceived for primary school children the books are very relevant to current work in secondary schools.

Gattegno, C., *For the teaching of mathematics*, vol. 3, Educational Explorers, 1964, 75p. A collection of essays on Cuisenaire rods; the author introduced the material to this country and developed its use extensively.

dominoes

Domino sets are available from most toyshops, but useful sets can be made with 3 cm × 6 cm cards marked with figures. A standard set D_6 might be described as a set of unordered pairs (a, b) where a and b are whole numbers from 0 to 6. Removing all the pairs that include a 6 leaves the set D_5. In general the set D_n will run from a double zero to a double n.

How many dominoes are there in D_6?
How many dominoes in D_n?
(1, 3, 6, 10, . . ., the triangular numbers).

How many dots altogether in a standard set D_6?
What is the sum of all the numbers in D_n?
(0, 3, 12, 30, . . ., n times the number of dominoes).

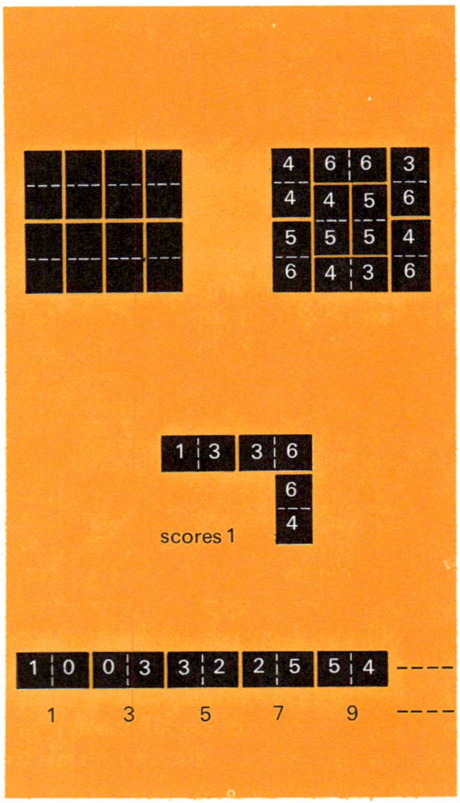

Arrange 8 pieces of a domino set to form a magic square of 16 numbers; i.e. one in which the sums of each row and column are the same.
Can you make other magic squares?

In the standard game of dominoes, chains of pieces are formed with two adjacent pieces having the same number.
In one variation of the game, each player tries to play so that the sum of the numbers at the open end is a multiple of 5. The players score a point for each 5 that they make in this way.
Make up some other variations.

Can you make a chain of dominoes so that the sums of the numbers on each domino are successive odd numbers. What is the longest such chain?
Consider chains of dominoes that produce other sets of numbers. Try other rules for deriving a sequence from a chain, e.g. take the difference of the numbers on each domino.

Can you make a closed chain using all the pieces of a standard set D_6?

Investigate closed chains with D_n.

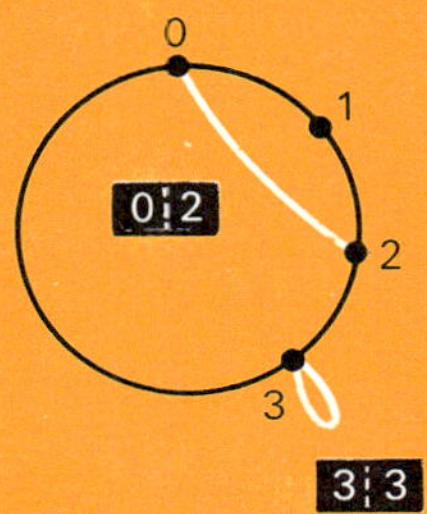

If $(n+1)$ points are marked on the circumference of a circle and numbered from 0 to n, then each domino of D_n can be represented by a line joining two points, the doubles being represented by loops at each point.

The problem of making closed chains with all the pieces of D_n can now be related to that of tracing a route over this network, passing once and only once over each line or loop and returning to the starting point.

For what values of n can such a route be traced?

Explain the following party trick: surreptitiously remove a domino, say (4, 5), from the standard set. Forecast that any chain formed from the set will have a 4 and a 5 at the open ends.

Where closed chains cannot be formed using all the pieces of D_n, what is the largest closed chain that can be made with some of the pieces?

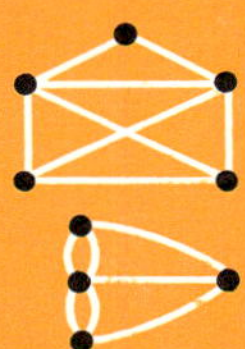

The tracing of so-called unicursal routes on any network is an interesting investigation in itself. In making up their own networks and tracing routes on them, children master the situation surprisingly quickly. The problem began with Euler's investigation of routes over the Königsberg bridges.

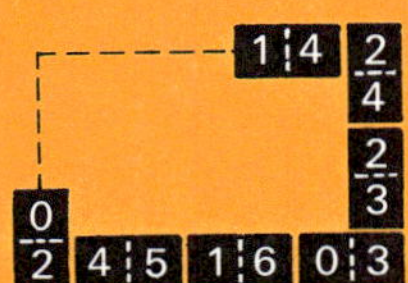

Consider making chains of dominoes such that the adjacent numbers on two consecutive pieces add to n. For what value of n can you make a closed chain using all the pieces of D_n with this rule?

(This variation in the touching rule has the interesting effect of reversing the previous situation. Where closed chains can be made with D_n using the first rule, they cannot be made using the second rule, and conversely.)

A different sort of domino set might be made by cutting out some triangles and marking the edges with, say, a 0, 1, 2.

How many different pieces will there be in the set T_2?

How many different pieces will there be in the set T_n?

Two pieces of T_n may be placed together touching along a side, if the numbers on this side are the same. What tilings can be made with this rule? What tilings can be made with other touching rules, e.g. that the numbers on the touching sides add up to n?

A variant of the triangular dominoes is available in the form of a game, *Contack*, made by Waddington. The triangular pieces of this game are coloured and numbered and suggest many challenging problems. Another game, *Hex*, made by Spears, uses hexagonal pieces whose edges are coloured in different ways.

Make a set of square dominoes marking each side with a 0, 1 or 2. How many different pieces will there be in this set S_2? How many different pieces will there be in a set S_n?

Make some square dominoes with colours instead of numbers. Select some of these to make various patterns, in particular those that cover the plane.

These extended domino sets can be fun to make and use to form patterns. To enumerate the different ways of colouring the edges of a regular polygon using up to a certain number of colours, can be difficult because of the symmetries involved. A general method of dealing with counting problems of this sort, due to Polya, is discussed in an interesting and elementary way by Golomb in his book on polyominoes (p. 171) to which the reader is referred in any case for another generalization of dominoes and a fruitful source for further investigations.

films

An increasing number of schools have projectors to show 8 mm film loops housed in special plastic cassettes. A cassette projector looking something like a television set costs about £100. This is simple to operate and can be used in moderate daylight. A smaller projector can also be held in the hands and projected on to wall or ceiling; this costs about £60.

The short 8 mm film loop, running time up to 4 min, has many possibilities. There are more than 200 currently available from various sources, costing from £1 to £4 each. These are listed in the A.T.M. film list (see below) which is revised from time to time.

It may be possible for a group of pupils to make short animated film loops themselves. The minimal requirements are:

(a) An 8 mm camera with a single shot facility. A zoom lens can be useful.

(b) A firm tripod or some home-made scaffolding made of wood or dexion strip.

(c) Lighting—say two or three reflectors with 250 watt photoflood bulbs.

(d) Film—a reel of 50 ft indoor colour film costs about £1·50, the price including the processing by the manufacturers. Empty plastic cassettes cost about 25p each, and are not too difficult to load with short lengths of film, say from 3 ft to 12 ft. Films can be sent to Rank (see p. 232) for loading, at about 40p a cassette.

(e) A corner of the classroom or stockroom needs to be undisturbed and available for a few days at a time.

(f) Coloured paper, geostrips, string, map pins, acetate sheet and other materials.

(g) Splicing equipment—easily applied prepared pieces of sellotape are available in cheap packs. A film editor on which a film can be turned by hand and viewed in a small screen is a necessity for serious work.

Beginners often attempt over-ambitious scripts. Initial experiments with short sequences of animations in different ways and different speeds are recommended. It is advisable to start with short sequences using simple elements.

For example two rings cut out of cardboard may be placed below the camera on a backing sheet of paper. The rings can then be moved a little at a time in a prescribed way. There is no need for a detailed accuracy in the movements—a nudge of $\frac{1}{2}$ cm or so every two clicks of the shutter will produce an adequately smooth motion though it will be obviously useful and desirable to experiment.

A film loop would appear to be most effective if the sequence ends as it begins. Thus the film mentioned on p. 97 would be only about a metre long, a matter of a 100 single animations. Further useful advice on making films can be found in an A.T.M. pamphlet listed below.

Seeing film could be a regular mathematical experience of a child. It is not always the case that they would be followed immediately by analysis or further investigation. But they can form a stock of experience which may be referred to or used later. Considerable sensitivity is required to judge how, when, and if further work arises directly from a film. In the case of the example given of a film being used as a starting point (p. 150) it is assumed that the reader will have understood that the timing involved is a matter of interpretation of the particular situation.

The need for sensitive timing is obviously important when the children are shown abstract film cartoons. Many of them appear to have interesting mathematical aspects that tempt us to forget that the film is its own end. Nevertheless we give some comments on such a film, not as a prescription but as a detailed illustration of the range of activities that might be taken from a film when the choice to do so is made.

Dance Squared is a 16 mm film cartoon, in colour and with sound, lasting about 4 minutes. It can be purchased from the National Film Board of Canada for £7, or hired from the British Film Institute for 35p. (For addresses, see p. 232). Many teachers have found that it is a never-failing delight to children, as well as a spur to their inventiveness.

After seeing the film children might paint—they might make pictures with gummed paper—black makes a good background,
they might make stories
they might talk.

how many shapes did you see?
what colours were there?
you saw a house in the film?
what shapes was it made of?

make your own house . . . ('windows' were sixteenths)
make some patterns

the film starts with a square bouncing around the screen—describe these movements with diagrams if you wish

how many squares can you find in this figure . . .?
how many different squares?
—different?

in the film the shapes moved round the screen in various ways—investigate ways of moving a square from one position to another.

a triangle 'points upwards' describe some ways in which you could move it so that it is pointing downwards . . .
split it up and put it together again if you wish

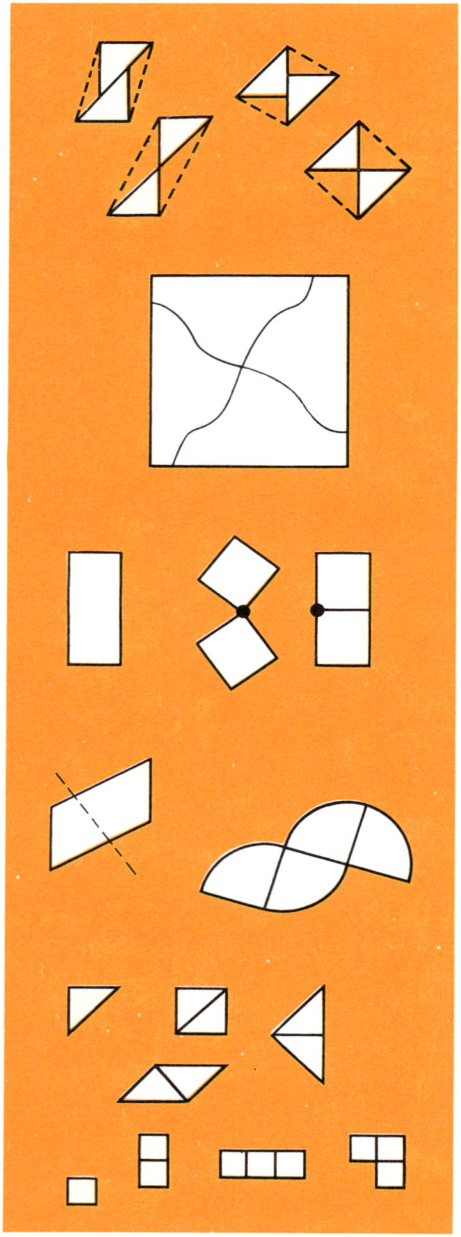

do you remember when squares and triangles are moving round in a circle? as part of the dance they reverse direction and move round in the opposite way—how is this done?

in the film there were two triangles sliding on each other—
imagine an elastic band round them as they move
draw some of the shapes the band would make.

split up a square into four equal parts—keeping any two of these fixed move the other two—draw or cut out some of the shapes you get (how many ways are there of splitting a square into four equal parts?)

in the film a shape is often broken up and reassembled in some way—can you describe some of these movements?
—diagrams?

make a flicker book

reassemble some shape in your own way. . . .
break up a parallelogram and reassemble it into a rectangle

break up a regular hexagon and reassemble it into a rectangle, into a . . .

can you reassemble a circle into another shape?

how many shapes can you make using 2, 3, 4 . . . squares triangles and other shapes?
—touching rules?
—polyominoes?

(Other recommended 16 mm film cartoons distributed by the National Film Board of Canada include *Boogie Doodle, Dots, Lines horizontal, Lines vertical, Loops, Marching the colours, Mosaic,* and *Notes on a triangle.*)

The classic 16 mm films by Nicolet on the geometry of the circle and other conics have been referred to on p. 175. These and other 16 mm films mentioned in the A.T.M. film list have to be shown in a room with blackout facilities, and a 16 mm film projector and screen. Many 16 mm films can be hired by the day at charges varying from under 50p to £3 depending on the running time. There are a few films available on free loan, mainly from various computer firms including I.B.M. (see p.232).

While considering animated motion, the reader should not forget the joys of making a flicker book. This can be made by drawing figures in slightly changing positions on successive pages of a small pad. When the pages are flicked through the fingers, the figures form a simple animation. Children delight in this and produce a wide range of sequences from exuberant fantasy to controlled illustration of mathematical themes. Finished flicker books provide interesting starting points for further investigation. Some of the situations explored with acetate sheet (see p. 158) might be explored, for example, in this way.

A 8 cm × 10 cm pad of about 40 pages seems to be a useful size. Job lots are often available from stationers at about 3p a pad.

Finally in this context we should remind the reader of the television set as a resource that may be available in the classroom. The B.B.C. and the independent networks produce various programmes for schools which are advertized in the usual way and in broadsheets and pamphlets sent directly to schools. The secondary school course, *Maths today*, broadcast by the B.B.C., has some associated literature and 8 mm film loops. Workcards associated with the primary school course, *Maths Workshop*, are available. These have plenty of material suitable for younger secondary pupils.

some books:
A.T.M., *List of mathematical films*, A.T.M., 1968.
A.T.M., *Films and film making*, A.T.M., 1967.
Gattegno, C., *For the teaching of mathematics*, vol. 2, Educational Explorers, 1964.
Gattegno, C., *Towards a visual culture*, Outerbridge, 1969.

games and puzzles

Various games and puzzles found in toyshops and stores provide suitable and interesting material for exploration. Finding out what can be done with the material may be its own end. But sometimes the gained insights into the different possibilities and limitations may be formalized in some way. This mathematisation then yields further solutions to certain problems, and in turn throws up further problems for investigation. Such activity is a genuine and accessible form of operational research. There are many useful hints in some of the books on mathematical recreations.

We have taken two materials, dominoes and playing cards, and given two sections to them (p. 188 and p. 220) but we could just as well have chosen another two. Here we give only a brief list of some recommended items. The addresses of firms mentioned are given on p. 232. Unless otherwise stated prices may be taken to be less than 50p per item.

Construction sets

There are many interesting sets made with pieces that interlock in some way. The pieces may be essentially linear, as in *Meccano*, or they may be plane or solid shapes. We mention various examples.

Milk straws joined by pipecleaners make an inexpensive and easy substitute for linear connections. *Geostrips* are discussed in another section (p. 202). Another useful set is *Construct-o-straws* (left), containing pliable plastic straws with variously spoked connectors, made by R. J. M. Exports at 75p a set.

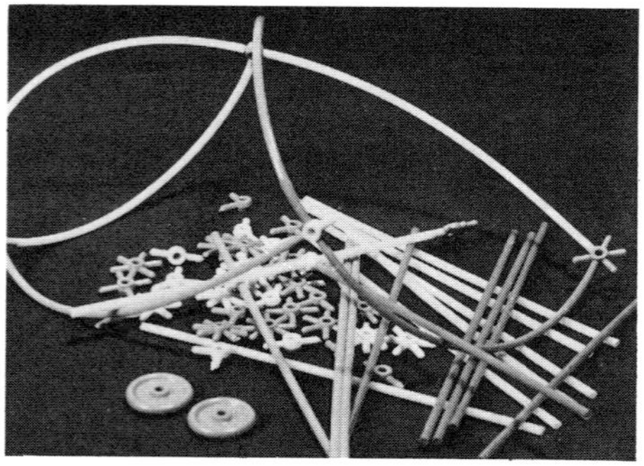

Attractive examples of plane connections are *Play-plax* (above), a set of coloured translucent squares and cylinders, made by Trendon, 50p a set, and *Octons*, coloured translucent octagons from Galt at 50p a set.

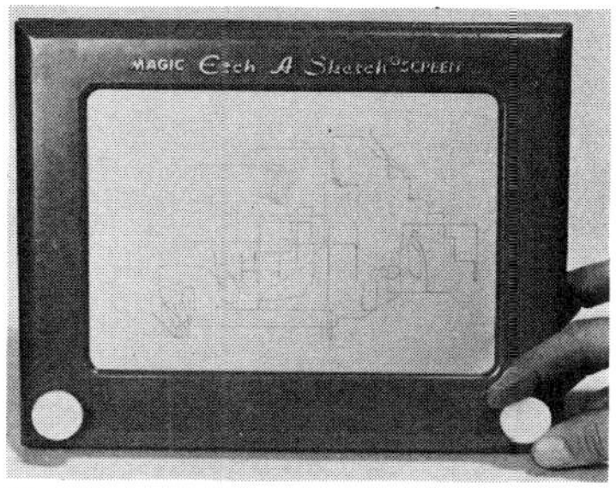

Designs

A delightful and very popular drawing kit is *Spirograph* made by Fisher at £1·75. Another recommended device is *Etch-a-sketch*, made by Fisher at about £2; this produces lines in two directions by turning two knobs.

Kaleidoscopes can also be mentioned in this context. Particularly recommended is one that has interchangeable heads, one of which can be filled with objects of your own choice, for example one made for Galt at 25p. A viewing type turns any scene into a kaleidoscopic pattern—from Bradley at about £1·50. Circular cards are rotated and seen through two mirrors in the *Specroscope* made by Spear at £1·00.

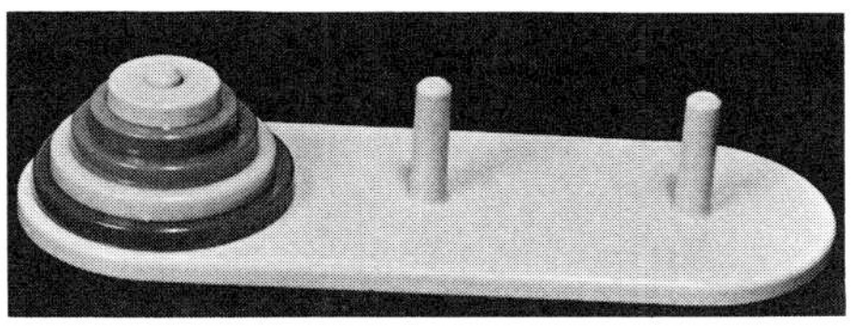

Permutation puzzles

The classic puzzle of this type is known as the *Fifteen puzzle*, and is usually available in inexpensive plastic form. Also recommended are *Tantalizing Seven* (Pan playthings) and *Move-the-Mountain* (Galt).

A famous puzzle known as the *Tower of Hanoi* is made in a convenient plastic form by Taskmaster.

Another toy that might be mentioned here is *Think-a-dot*, a computer toy discussed on p. 146.

Pegboard games

Many well known games using counters, chess-board and so on may be played with pegs on pegboard. The A.T.M. pamphlet no. 13 lists some of these, with some hints for mathematisation. Pegboard is also discussed on p. 216.

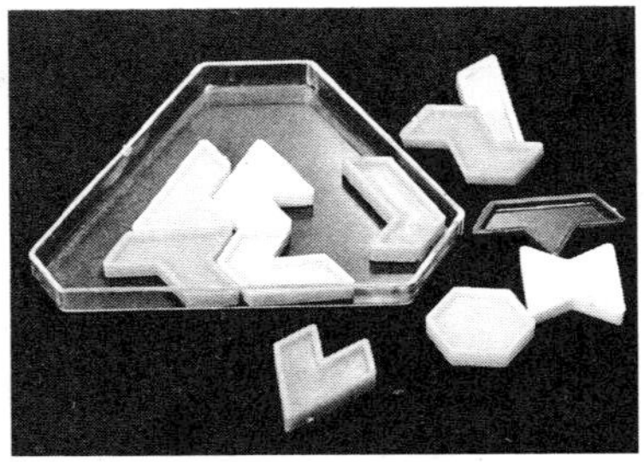

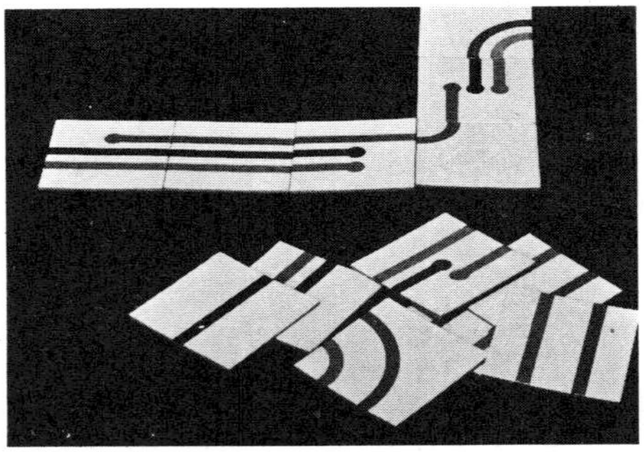

Tiling puzzles

For some discussion of tiles in general see p. 226.

Various inexpensive dissection puzzles, attractively made in coloured plastic can be found in toyshops, e.g. *Enigma*, a dissection puzzle with various shapes, and *Maestro*, a pentomino set (both made by Pan Playthings); also *Multi-puzzle*, a set of hexominoes made by Spear.

There are various domino-type games using coloured cards that are tiled in some way. In particular, we mention those that make network patterns such as *Lin-Jo* (Morton, 50p) and *Connect* (Galt, 50p).

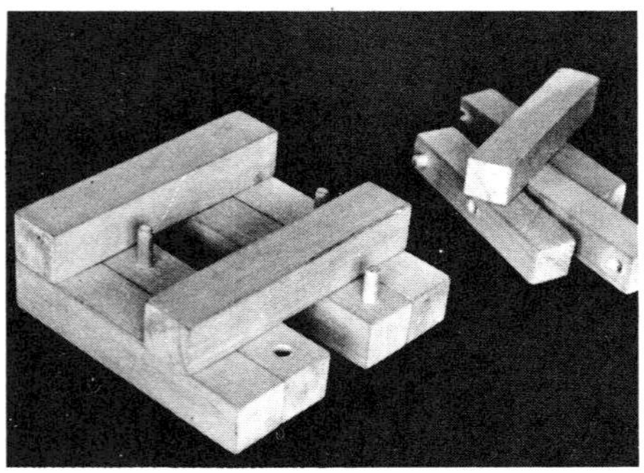

Three dimensional puzzles, such as *Soma-cubes* and *Cube-it* have been mentioned elsewhere (p. 181). An intriguing and unusual *puzzle* with 10 wooden rods, each of which has one protruding dowel and one hole, all to be assembled into an interlocking block, is made by Abbey Products at 65p.

Traditional games

To complete this brief selection, this heading serves as a reminder of such things as playing cards, dominoes, dice, draughts, solitaire, paper-and-pencil games and so on, some of which are discussed in more detail in other sections.

some books:

A.T.M., *List of materials for mathematics*, A.T.M., 1968.

A.T.M., *Pegboard games*, A.T.M., 1968.

Ball, W. W. R., *Mathematical recreations*, Macmillan, 1949.

Bell, R. C., *Board and table games*, 2 vols., O.U.P., 1969.

Cundy, H. M. & Rollett, A. P., *Mathematical models*, O.U.P., 1961.

Domoryad, A. P., *Mathematical games*, Pergamon, 1964.

Gardner, M., *Mathematical puzzles and diversions*, Bell, 1961.

Gardner, M., *More mathematical puzzles . . .*, Bell, 1963.

Golomb, S., *Polyominoes*, Allen and Unwin, 1966.

Kraitchik, M., *Mathematical recreations*, Allen & Unwin, 1943.

Lucas, E., *Recréations mathématiques*, 4 vols., Blanchard, 1960.

Sackson, S., *A gamut of games*, Random House, 1969.

geoboards

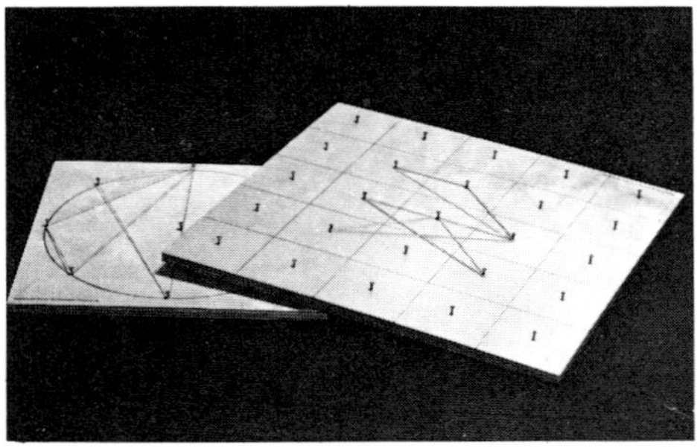

These are boards with a lattice of nails on which various shapes can be made with elastic bands. 9-, 16-, and 25-pin square lattice boards, and 8-, 10-, and 12-pin regular polygon boards are available from Cuisenaire at 55p to 70p each. Home made versions can be made with thick plywood or chipboard and escutcheon pins (about 12 mm) from ironmongers. For the square lattices, nails should be about 5 cm apart; Cuisenaire supply useful squares of adhesive plastic already marked out, a set of four costing 90p. Good strong coloured elastic bands are sold by some educational suppliers and may be preferable to the ones usually sold by stationers which soon snap.

The geoboard is a versatile and invaluable material and there is a wide range of activities for which it may be used. We have given three starting points using geoboards on pp. 70, 78 and 105, though these only exploit some possibilities in a casual way. We list a few sample activities here and refer the reader to the extensive accounts in the books listed below. We would urge the reader to have geoboards easily available in the classroom. There is nothing quite like the concentration with which people play with them.

Make patterns.

Make some specific shapes.

Make triangles.

How many different triangles can you make? Can you make an equilateral triangle?

Choose a unit of area and find out the area of various shapes you can make on the board.

What about the areas of shapes made by bands intersecting but not at nails of the lattice?

Explore some of the situations in the sections on grids (p. 206) and pegboards (p. 216) using a geoboard.

Make a unit triangle on the board. The triangle may be moved to another part of the board, say by making two shifts to the right and one shift down. The triangle is now in a different position on the board, but the shifts and combinations of shifts have preserved size, shape, and orientation.

Investigate other transformations, and consider what is preserved in each case.

The following game was invented by two pupils, as a result of such an investigation.

The rules are:

(a) The two players start in opposite corners (preferably the top left corner and the bottom right corner). The person whose rubber band is on the top left corner starts first and thereafter turns are taken.

(b) Each shape formed by the rubber band must have four sides.

(c) The area must be preserved.

(d) A positive move must be made, and at least one point must stay the same.

(e) The player must move either (i) towards the other player's rubber band, or (ii) to leave his band the same distance away from the other player's band as in the previous move.

(f) To win, three out of the four points of the opponent's band must be covered.

It is perhaps just worth pointing out another use for the elastic bands when slipped round a rectangular card. A set of rectangles, say about 8 cm × 10 cm, made from cardboard or hardboard is required. The edges of the card should be roughened.

Slip one elastic band round the card.
What are the various possibilities?
Slip two bands round the card.
Slip three bands—and so on.

some books:

A.T.M., *Notes on mathematics in primary schools*, C.U.P., 1967.

Gattegno, C., *For the teaching of mathematics*, Ed. Explorers, 1963.

Gattegno, C., *Geoboard geometry*, Cuisenaire, 1968.

geostrips

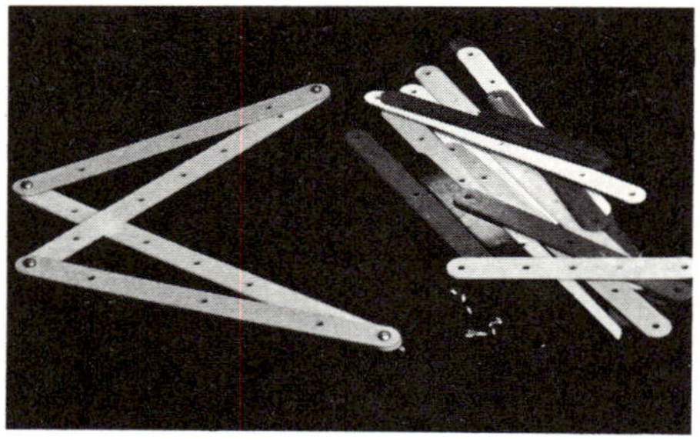

Geostrips are coloured plastic strips of various lengths from about 30 cm to 8 cm. The strips are punched at each end and at intervals along the length so that two strips may be linked in various ways with split-pins. A set, containing six dozen strips, some fasteners and a protractor, costs £2·85 from Taskmaster (see p. 233). The attractive colouring and the simple method of linking makes this pleasant material to play with and pupils should be encouraged to explore it before any effort is made to direct their activities.

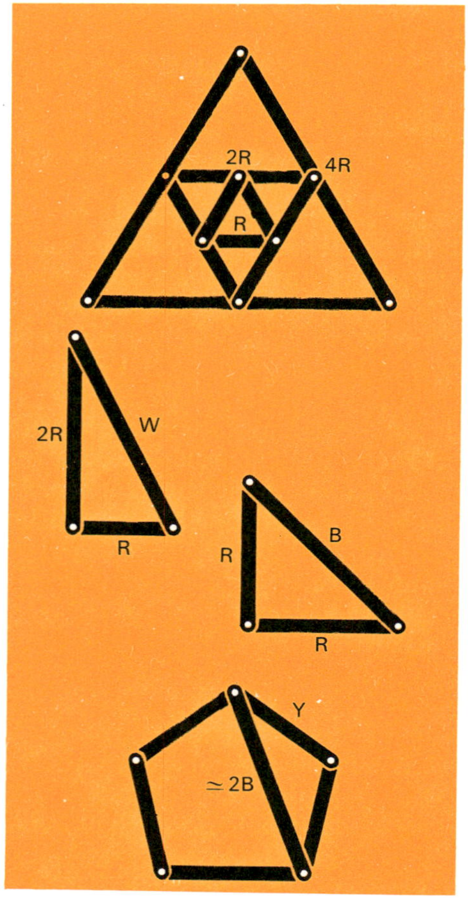

Very soon it will be realized that each colour (except white) is related within its own set, i.e. the small red strip (R) is half the medium red strip (2R) and this is half the large red strip (4R). Thus the various lengths might be referred to like this: R, 2R, 4R; Y, 2Y; B, 2B; W; W′.

The relationship of colours can be sought through constructions of triangles and other polygons. Right-angled triangles seem a fruitful way of doing this. Taking R as a unit, the right-angled triangles that can be made relate all the strips except the longer white one, W′.

The triangle made with the two white strips and the small blue one looks right-angled. This suggests that the larger white strip, W′, might be $\sqrt{7}$R. But in fact W′ is a diagonal of a regular octagon with side R. The play in the linkages, in a sense allows both possibilities. Again, 2B, which is accurately the diagonal of a square of side 2R, also serves as the diagonal of a pentagon of side Y. The investigation of these relationships is interesting and rewarding.

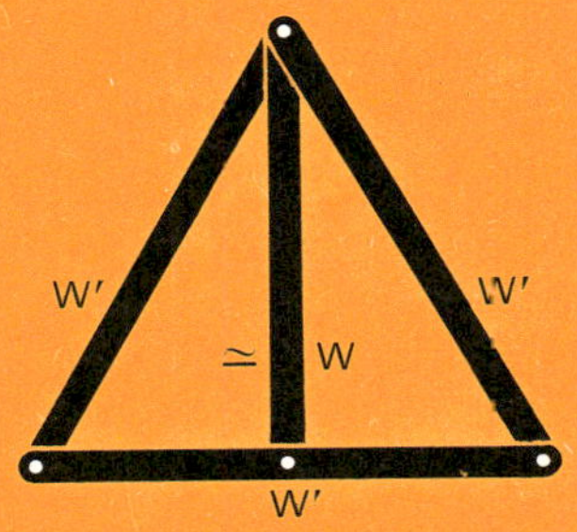

Some concentration on different kinds of triangles and the inclusion of altitudes brings into use the extra holes that are punched on some strips. This will give an approximate link between W and W′.

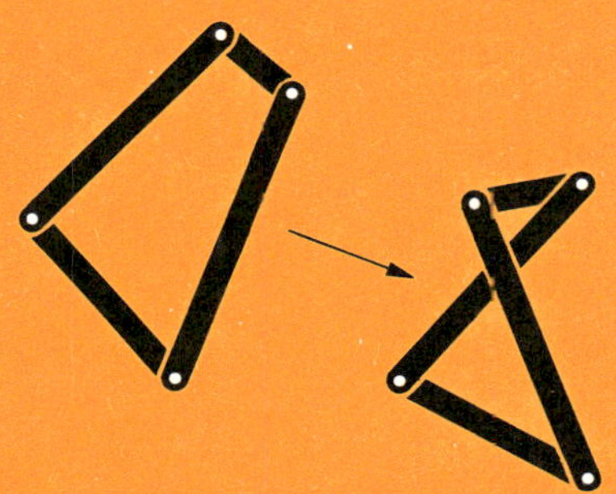

It is during the early period of exploration that pupils realize that the shapes being made have various degrees of freedom—the triangles will not move, the quadrilateral takes many forms. . . .

What is a quadrilateral?

What happens to its diagonals?

Take any 4 strips—make a quadrilateral. This will produce a variety of quadrilaterals from a group of pupils. Some discussion of shapes with strips all of the same length, 3 of the same length, 2 of the same length . . . will identify the different quadrilaterals and emphasize the special cases of parallelograms (rectangles) and rhombuses (squares).

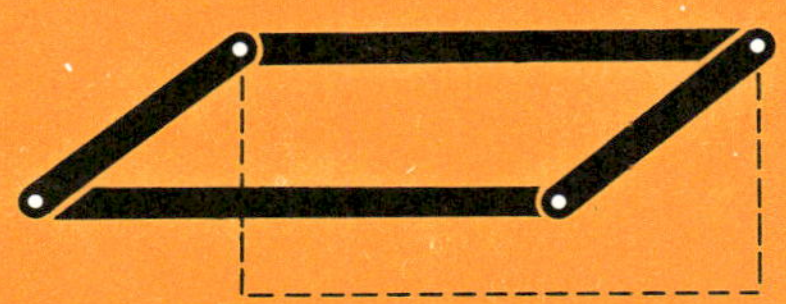

Consider all the possibilities with only 3 strips.

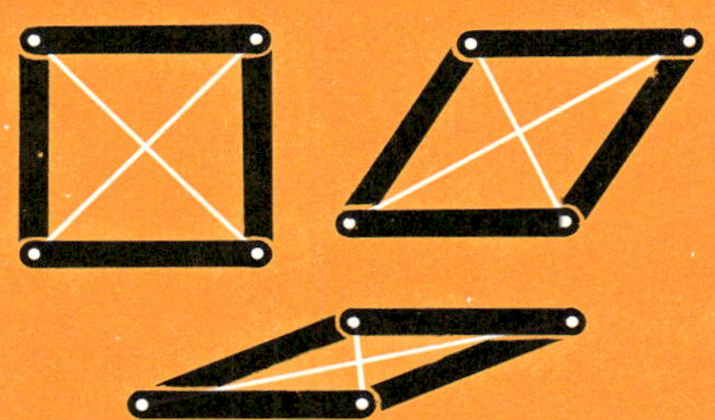

Take one of the quadrilaterals and make diagonals with shearing elastic. Move the quadrilateral—what happens to the diagonals? Make diagonals for other quadrilaterals.

When are the diagonals equal?

When do they cross at right angles?

When is one half the other?

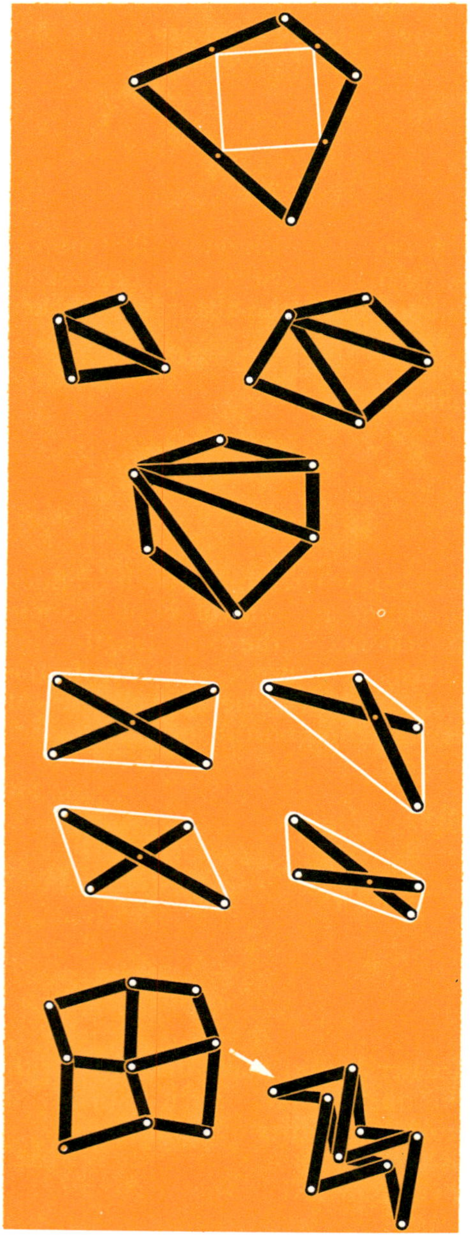

On one of the quadrilaterals join some of the spare holes with shearing elastic. What happens to the 'elastic' shape when the quadrilateral is moved?

Try this with other quadrilaterals.

Any triangles you have made cannot be moved in the same way as the quadrilaterals. The triangle is rigid. Insert diagonal strips in the quadrilaterals to make them rigid—how many are required?

How many diagonals are required to make other polygons rigid? Investigate this for different polygons to produce a function relating sides (s) to numbers of diagonals (d); sides (s) to triangles (t); sides (s) to angles (a).

Make a cross with a pair of strips. Join their ends with shearing elastic to form a quadrilateral.

What shapes can be made with:

- equal diagonals joined in the middle?
- equal diagonals joined elsewhere?
- unequal diagonals joined in the middle?
- unequal diagonals joined elsewhere?

Use a number of strips of the same length to make a trellis. How long is it? How high is it? Move the trellis and investigate changing lengths and heights.

Make a trellis using strips of two different colours.

What effect does movement of the trellis have on the basic unit?

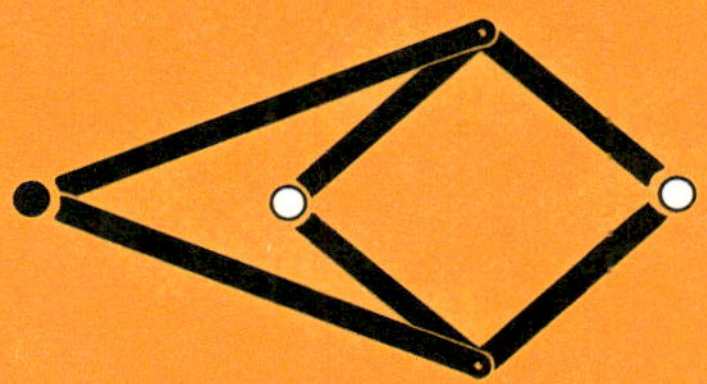

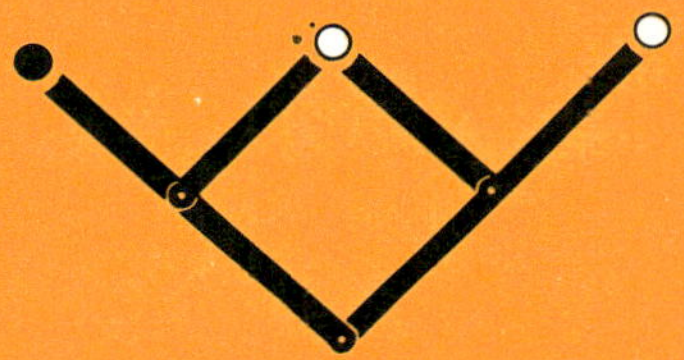

Almost any shape that is made with the strips has moving parts. It is this movement that appeals to the young who can make 'moving patterns' by simply holding some part of their shape as they move and draw around some other part. If the drawing is done with felt pens attractive patterns emerge.

Some of the drawings that follow show more directed and systematic linkages.

To make a pattern some parts of the linkage need to be fixed. These fixed points are shaded in the diagrams. The pattern is developed either by marking with a felt pen the positions of moving points, shown unshaded, or the different positions of a complete shape.

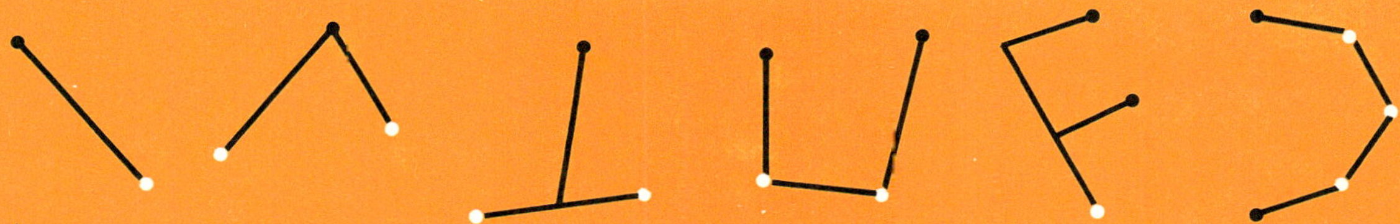

A particular linkage that can be used to make an enlargement of a given figure is called a pantograph. Find out how this works and explain what properties of the transformation it exhibits.

Another linkage, called a Peaucellier linkage after its inventor, may be used to investigate another geometrical transformation, inversion. In this case a line may be transformed into a circle and conversely (if one of the unshaded points moves in a straight line the other moves in a circle).

Investigate the properties of inversion by experimenting with this linkage.

Invent your own linkage and give an account of what can be done with it.

Further work may be found in *Machines and mechanisms*, a Mathematics for the Majority Project guide published by Chatto and Windus.

grids and lattices

In general the network of lines formed by the edges of tiles in a tiling may be called a grid. The corresponding pattern of points formed by the vertices may be called a lattice.

Paper printed with various grids and lattices is an essential material for mathematics. Square and triangle grids are available in various sizes. The 1 cm (or $\frac{1}{4}$ in.) grids could be used for most work. This includes the drawing of most graphs, so that the miserable 1 mm (or $\frac{1}{10}$ in.) grid could be used more sparingly. In fact most work with coordinates is more meaningfully done on a square lattice of points—or in the early stages with pegs and pegboard.

A range of useful grids and lattices are sold by Copyprint (see p. 232) at about £2 for 500 sheets (20 cm $\times$ 30 cm). A highly recommended set of five irregular grids is available from Altair Designs at £2 for 250 sheets—50 of each design.

Various sheets can be regularly duplicated from master stencils, and probably this is the cheapest and the most convenient way of keeping a supply of point-lattices and such special grids as the tiling of irregular triangles and quadrilaterals.

A brief account cannot hope to be exhaustive. Various activities described in other sections can clearly also be pursued with this material.

Take a particular grid or lattice.
Study it and let the imagination roam.
Mark in lines. Join points. Colour regions.

. . . suddenly it happens that someone is trying to find 4 points that make a square. Is this possible on a hexagonal lattice? A triangular one? A square one? What about other regular polygons?

Another person is making parallelograms that enclose 1 point, 2 points, and so on.

What rectangles can be found?

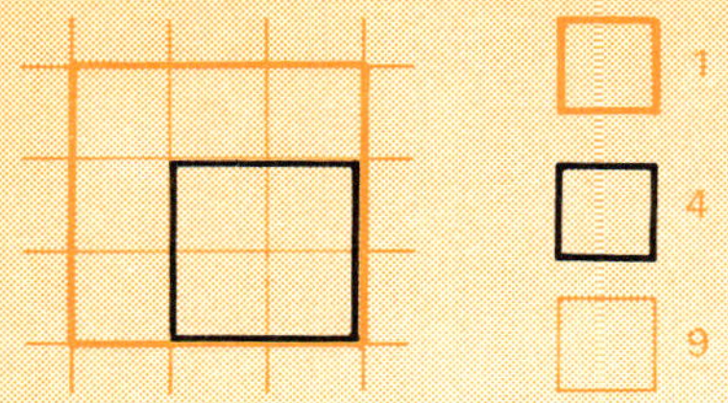

Mark a large square on a square grid.

How many squares can you see already marked in this figure?

Where the figure has, say, $n \times n$ small squares of the grid the answer to the question is the sum of the first n squares. What about rectangles? Surprisingly the answer is the sum of the first n cubes.

If similar questions were posed for a square lattice then it may be decided to count the oblique squares formed by lattice points of the figure. What is the general formula in this case?

Choose 3 points of a lattice.

What is the area of the corresponding triangle? What unit area did you choose?

What areas can you make with triangles on various lattices?

Make areas with polygons having vertices at lattice points. Count the number of lattice points on the boundary of each polygon, say b. Count the number of points inside the polygon, say i. Shapes with the same number b and the same number i turn out to have the same areas.

On a square lattice the area is $(\frac{1}{2}b+i-1)$ unit squares, (Pick's theorem). Are there corresponding formulas for the triangular lattice or the hexagonal lattice? (See p. 70).

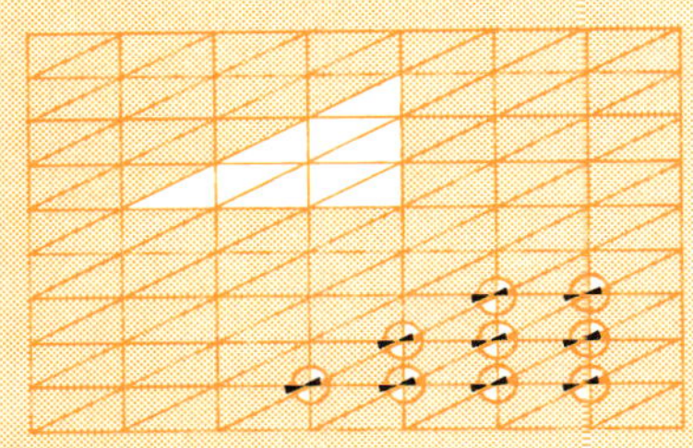

Consider the scalene triangle grid.
Colour one triangle.
Colour the next largest, the next . . .
How many unit triangles in each?

Mark an angle with a coloured pencil.

Mark all other angles of the same size with the same colour. Mark another angle in a different colour. Mark all others of the same size in the same colour.

Repeat this for the remaining angles.

Consider the angles at a point, the angles of various polygons, the angles between transversals and parallel lines.

Draw some circles on a particular lattice. How many lattice points are there inside each circle? Can you find circles that enclose 1, 2, 3 . . . points for various lattices?

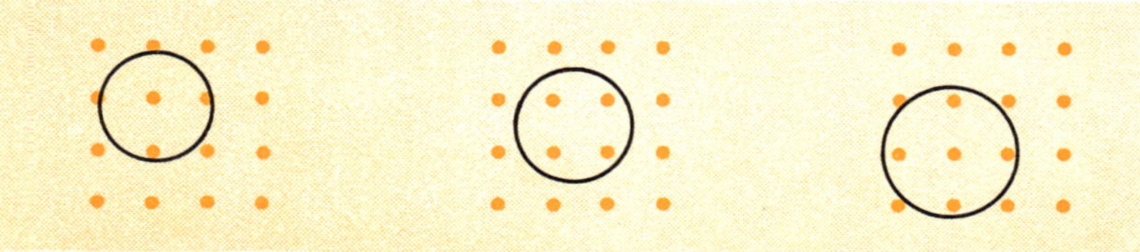

Suppose the circles are restricted to those whose centres lie at lattice points. Can you find circles that enclose 1, 2, 3 . . . points in this case?

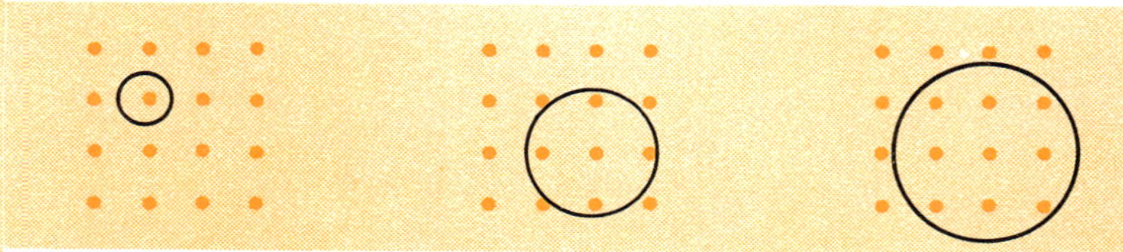

For a square lattice, the number of points inside or on a circle counts squares that are in, or partially in, the circle, so that this number gives approximately the area of the circle. This may be used to find approximations for π. A circle with centre at a lattice point, and radius 10—the unit being the distance between lattice points—gives $\pi = 3{\cdot}17$. A circle with radius 100 gives $\pi = 3{\cdot}1417$.

Mark as many points of a square lattice as you can that are a distance of 2, 3, 4 . . . units from base point. These may be found directly or by making circles with centres at the base point. For what distances will there be such points not on the axes through the centre?

For example, the point (3, 4) is 5 units from the centre. The triplet (3, 4, 5) has the property that $3^2+4^2=5^2$. What about the other triplets derived in this way?

The reader will recognize some arithmetical aspects of Pythagoras' theorem. The investigation of whole numbers which can be the hypotenuse of a Pythagorean triplet is related to the problem of expressing a number as the sum of two squares. Thus 5 is a hypotenuse of a Pythagorean triplet and $5=1^2+2^2$. But there is no triplet with hypotenuse 7, and 7 cannot be expressed as the sum of two integral squares.

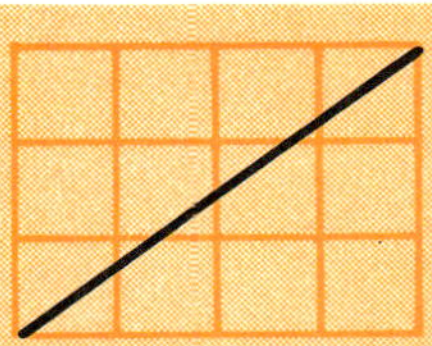

Draw a rectangle on the lines of a square grid. How many squares does a diagonal of the rectangle pass through? (See p. 6). Explore some corresponding problems on a triangular grid.

Draw a rectangle with vertices at the points of a square lattice. Starting at one corner draw a diagonal line as shown until it meets a side of the rectangle. Rebound at right angles off this side. Continue the path in this way. Which paths are formed for various rectangles? How long is the path in each case? Where does the path end?

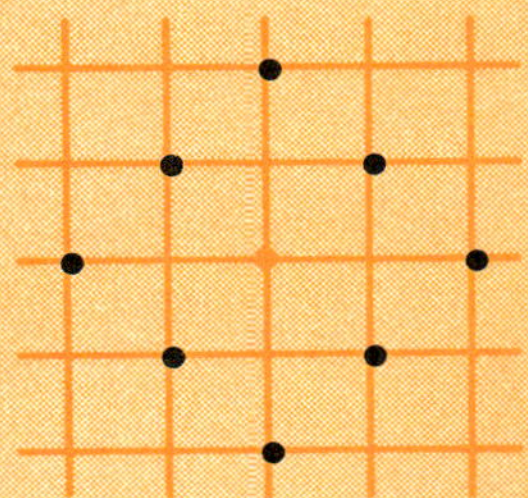

On each grid an appropriate notion of distance can be defined in terms of routes along grid lines. Thus the points shown are at a distance of 2 units from the centre—and this distance is the shortest distance in each case. The set of eight points shown defines a 'circle of radius 2' in this geometry.

Explore other ideas in this geometry, e.g. perpendicular bisectors, mid points, ellipses, squares and so on. What theorems can you find?

Consider similar problems in the geometry of other grids.

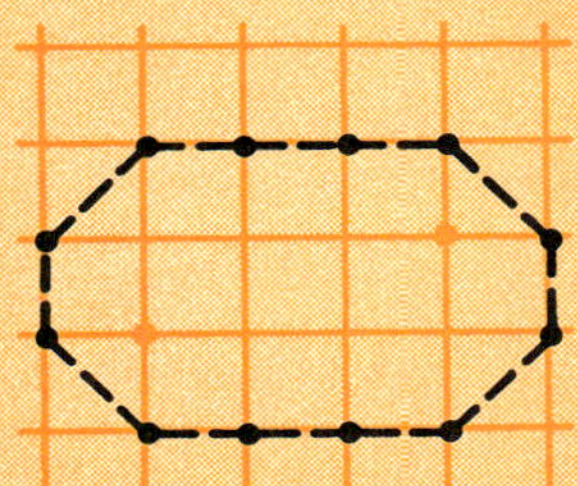

'ellipse' with two foci

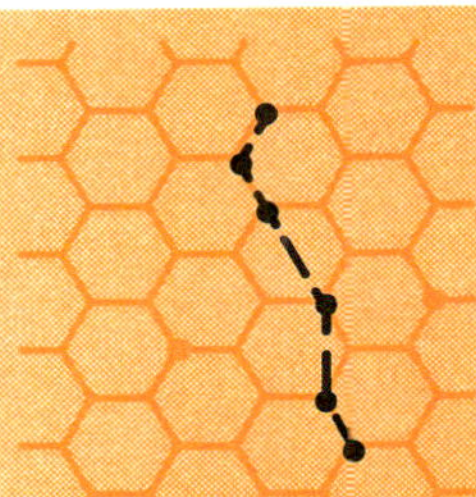

'perpendicular bisector'

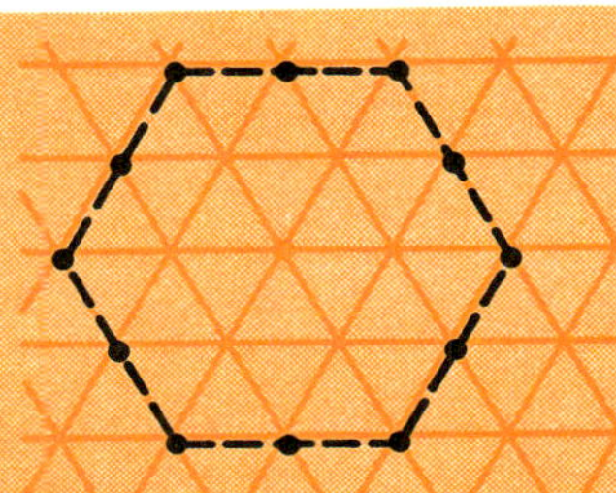

'circle' of radius two

How many shapes can be made on the regular triangular grid that are made up of 2 triangles, 3 triangles and so on? Consider the corresponding problem on a hexagonal grid.

On a square grid shapes made up of 3 squares, 4 squares and so on are called polyominoes. (See pp, 118, 231.)

Some investigations of routes on a square grid have been mentioned on p. 130.

A wide range of further activities are possible both on the square and other grids. As an example, consider taking unrestricted routes on a hexagonal grid. On arriving at any junction along a grid line there are two possible choices of direction, assuming that the line along which we arrived cannot be chosen. If we code the two choices as L, R, for left and right turns, then any route on the grid can be described by a 'word', i.e. a sequence of letters L or R, e.g. LRRL which could be contracted to LR^2L.

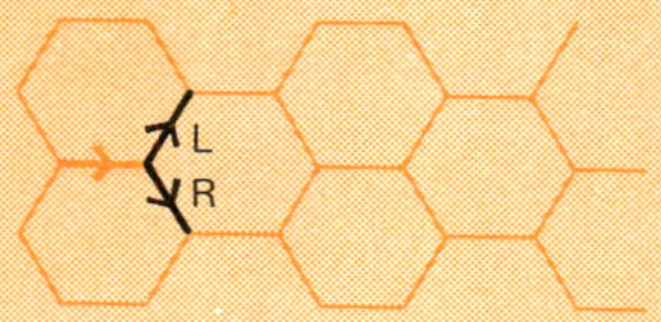

Conversely any such sequence will code a particular route. Shall LR mean go right and then left, or conversely? It is a matter of choice. In either case, LR is not the same as RL.

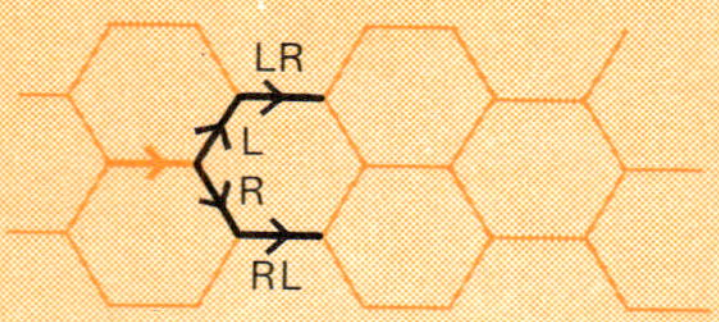

Note that the initial direction is always fixed. Also that we are labelling lines of the figure, not points. Thus LR^2, namely LRR, and R^2L, namely RRL, get to the same point but from different directions. To establish an equivalence of words a further step must be taken along each route, giving $RL^2R = LR^2L$. Find other relations of this sort, e.g. $LR^6 = L$.

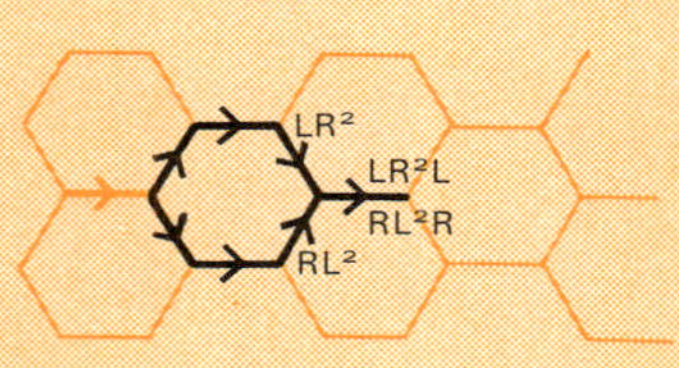

Corresponding to each word is a chain of equilateral triangles. Investigate the shapes made for some particular words.

Consider chains of other tiles and the corresponding routes on the appropriate network.

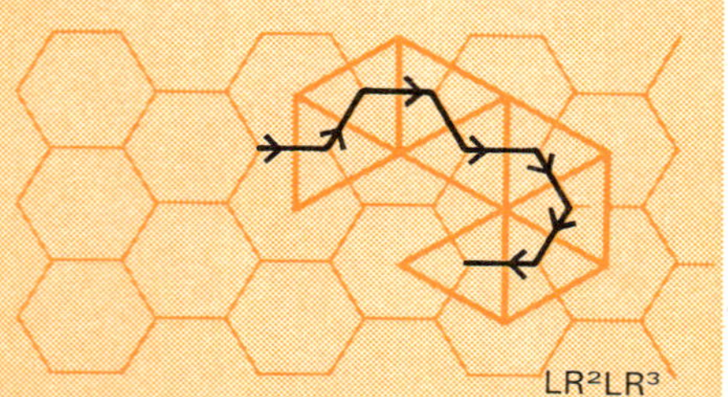

Investigate an algebra of 'words' in three letters derived from routes on a hexagonal grid in which three directions can be chosen at each point.

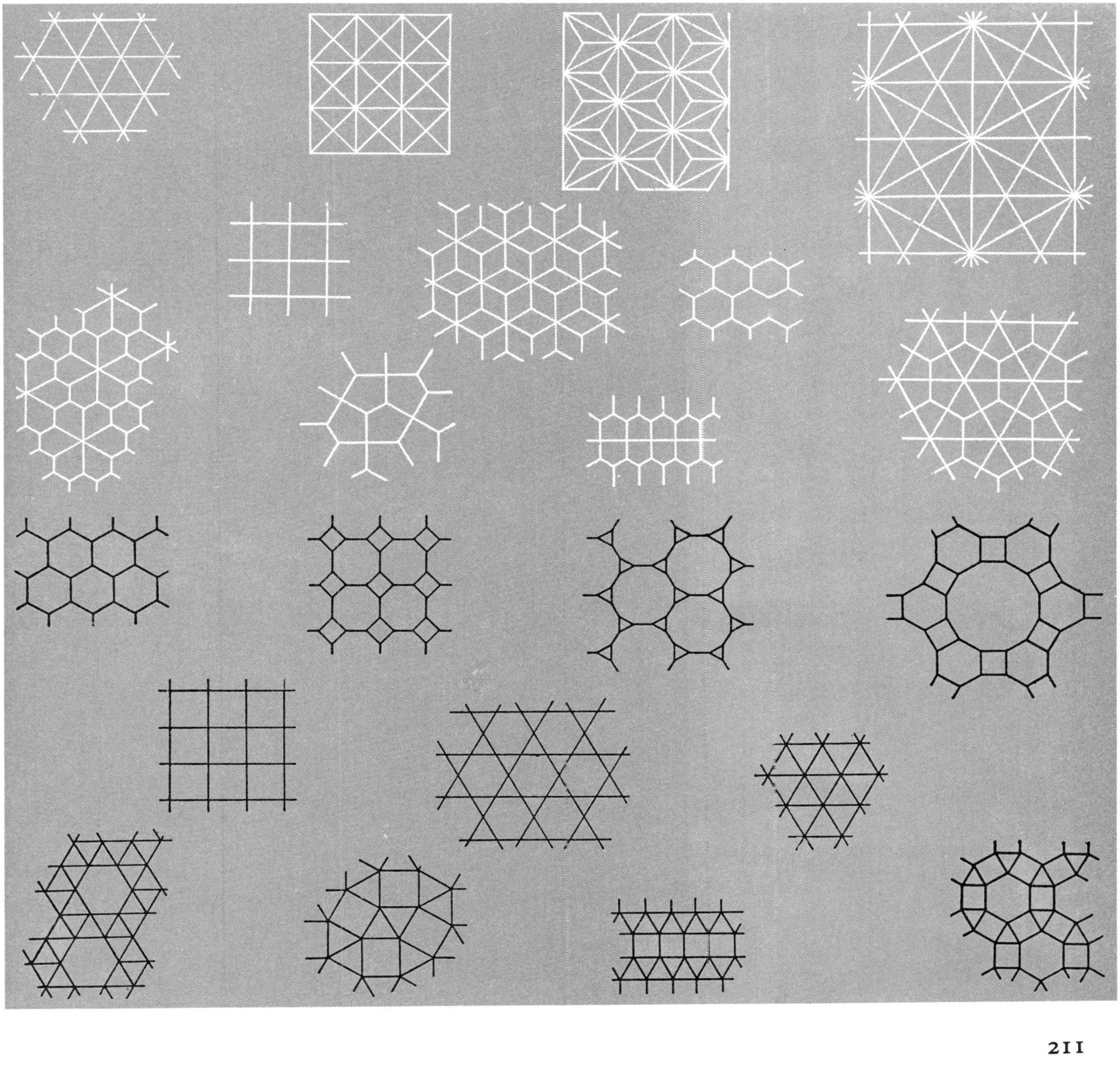

paper and card

Almost any kind of plain paper is suitable for folding activities and the paper used for duplicating machines is probably the cheapest source. The best results are obtained with cartridge paper. Paper of this thickness is recommended for making the skeleton models referred to below. Circular filter papers are also useful and can be obtained in various sizes from most chemists. Paper with gummed back is useful for recording work on dissecting shapes. A good source is Charles Ltd (see p. 232).

For making polyhedra a good supply of 6-sheet card is essential. This can be purchased in imperial size in various colours. The coloured card does not show grubby marks as easily as white.

Thin card can be cut with scissors but cutting knives and metal straight edges are necessary for thicker card. A guillotine is a valuable piece of equipment for the mathematics department.

Any quick drying glue such as balsa cement or Cow gum is satisfactory for sticking paper and card. Paper fasteners of the 'push through–bend back' type are required for making linkages (a punch is useful for making holes for these fasteners).

models

For instructions on making polyhedra from nets see the book by Cundy and Rollet listed below. Thin card, (6-sheet), is suitable material for constructing such models and flaps may be fastened together with a quick drying glue such as balsa cement.

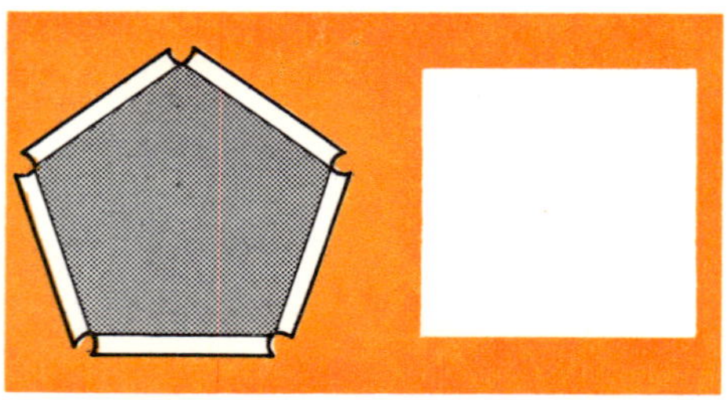

An alternative way of constructing regular polyhedra is to cut out individual faces, using templates, which have flaps attached at each edge. A small hole is cut at each corner—a hole punch is useful for this. The flaps must be scored and bent outwards. Faces may then be joined together by placing elastic bands around the flaps, thus avoiding the use of glue.

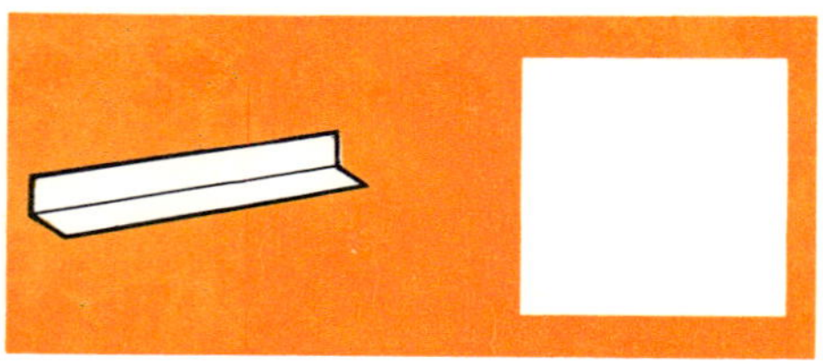

Skeletons of polyhedra may be made from cartridge paper in the following way: use strips of paper about 2 cm by 15 cm. With a ball-point pen firmly mark a line down the middle of each strip and fold along this line. The strips are fastened together with a quick drying glue and the models formed are surprisingly rigid.

folds

Many angle and line properties of geometric figures may be demonstrated by folding.

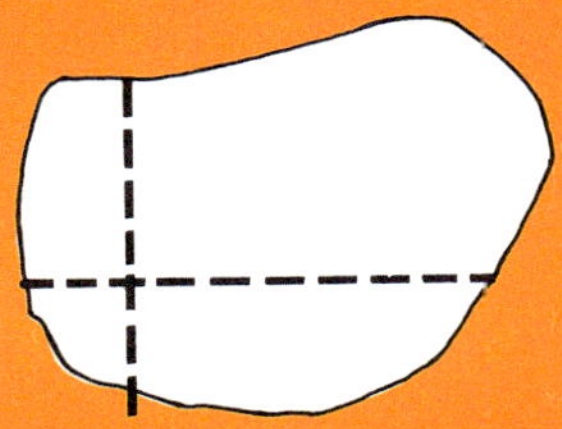

By folding an irregular piece of paper form a right angle, a square.

By folding a rectangular piece of paper form an equilateral triangle.

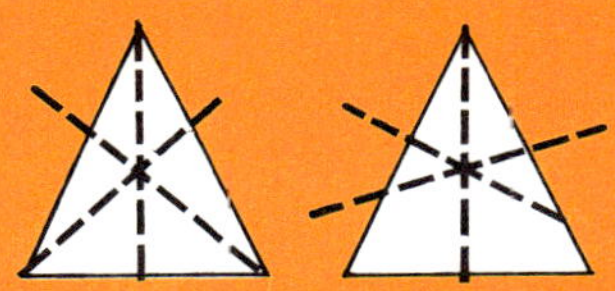

The bisectors of angles and the perpendicular bisectors of straight lines are easily obtained.

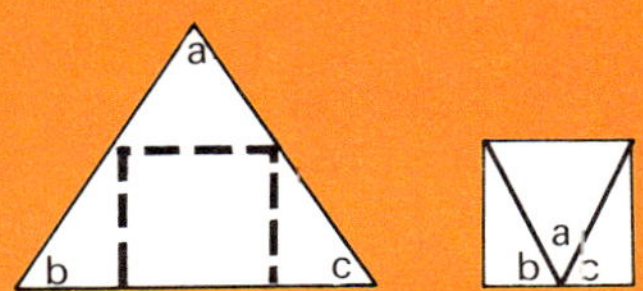

A triangle can be folded to show that the sum of its angles is 180°.

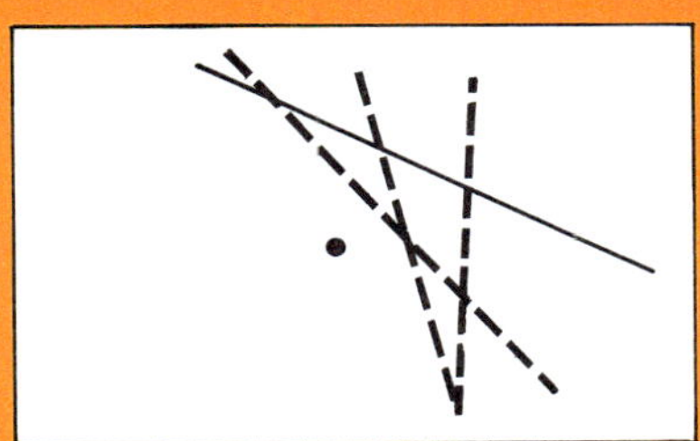

A series of folds may produce an envelope of a curve (for such exercises tracing paper is essential).

For example, draw a line and mark a point on a piece of paper. Fold so that the point lies on the line. This can be done repeatedly giving a series of folds which form an envelope of a parabola.

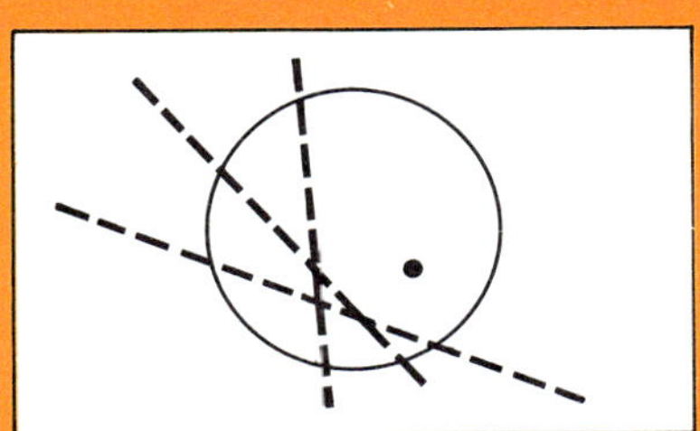

Draw a circle and mark a point inside it. Fold so that the point lies on the circumference. This time the series of folds form an envelope of an ellipse.

What happens if the point is at the centre of the circle? or outside the circle?

dissections

The delightful films *Dance Squared* and *Notes on a Triangle* are based on dissections of a square and an equilateral triangle. Seeing these films might well lead to interesting work on the dissection of polygons (see p. 193).

In how many ways can a square be cut in half by a single straight cut?

What shapes can be formed by dissecting a regular hexagon and reassembling the pieces?

Can any regular polygon be dissected and reformed into a parallelogram?

cut-outs

Fold a piece of paper. A shape cut out across the folded edge will have a line of symmetry. What happens when the paper is folded twice? Three times?

Which letters of the alphabet can be cut from a folded piece of paper?

linkages

These can be constructed from strips of thick card (about 10-sheet) joined by paper fasteners. A pencil or ball-point pen is placed through a hole in the linkage and traces out a locus as the linkage is moved. Investigate what happens as the lengths of the strips are varied; as the distance between the fixed points is varied. Cf. p. 205.

See the book by Cundy and Rollett for further work.

bands

A narrow strip of paper with a line marked along the centre is given a twist and the ends are fastened to give a so-called Moebius strip. The resulting surface has only one side.

What happens if the strip is given 2 twists or 3 twists before the ends are fastened?

In what other ways can pieces of paper be twisted and fastened? (see p. 144).

knots

Tie a knot in a strip of paper. Flatten carefully and the knot will form a regular pentagon.

Can you make other regular polygons by knotting paper?

In some cases, for example the regular hexagon, two strips of paper have to be knotted together.

hexaflexagons

A strip of paper on which are marked 19 equilateral triangles is folded in a special way to form a regular hexagon which may be *flexed* to show different faces. See the book by Gardner listed below for instructions.

plaiting

A net is drawn on paper, cut out and folded in a certain way to make a polyhedron. No sticking is necessary, the model being held together by the folded paper.

some books:

Cundy, H. M. & Rollett, A. P. *Mathematical models*, O.U.P., 1960.

Gardner, M., *More mathematical puzzles and diversions*, Penguin, 1966.

Johnson, D., *Paper folding*, N.C.T.M., 1957.

Mold, J., *Topics from mathematics; Solid models*, C.U.P., 1969.

Paling, D., Banwell, C., Saunders, K., *Making mathematics, Topic book: Making models*, O.U.P., 1970.

pegboard

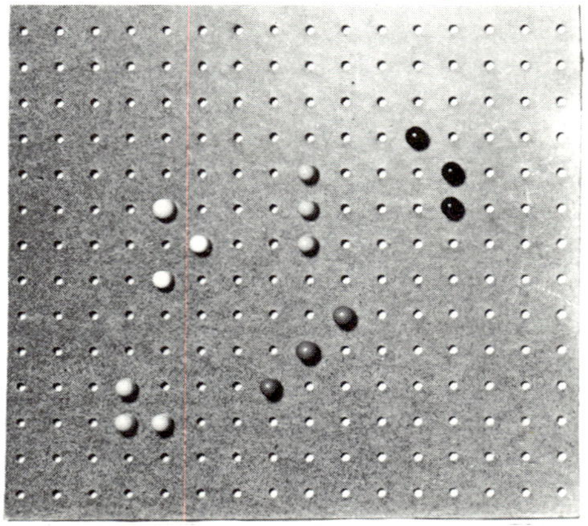

Holed hardboard is often available in cheap offcuts from do-it-yourself shops. 30 cm × 30 cm square pieces are recommended, with a liberal supply of pegs in various colours available at about 40p a 100 from the educational suppliers. Pegs are made in slightly different sizes, and it is worth checking that those ordered fit the standard pegboard holes. Placing pegs in the holes is pleasurable, and plenty of opportunity should be provided for undirected play with the material. The holes of standard pegboard form a square lattice and clearly many of the relevant activities may be followed with squared paper. (See p. 206.)

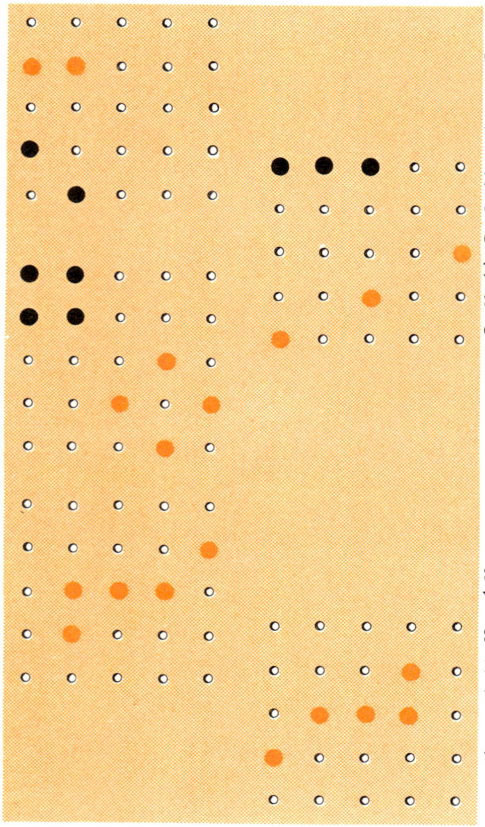

Various patterns can be made with pegs based on colour or on the position on the board. What constitutes a pattern? Can you make a pattern with 1 peg? With 2 pegs? With 3 pegs? Can you make a random pattern?

Usually an arrangement is felt to be a pattern when some rules are formulated describing how the pegs are placed, e.g. 2 pegs next to each other, 3 pegs in a line, 4 pegs at the corners of a square. Such rules may be ambiguous in the first instance. Making *different* patterns often forces a sharper formulation. Counting different patterns involves deciding when two patterns are the *same*.

Certain movements of the first board shown will bring the shape on it into the position of the shape on the other board. What movements? We may decide to ignore the effects of such movements in which case we would count the arrangement as giving only one pattern. If we do not ignore them there will be two patterns.

Any 'how many' question raises fundamental mathematical issues.

How many . . .?

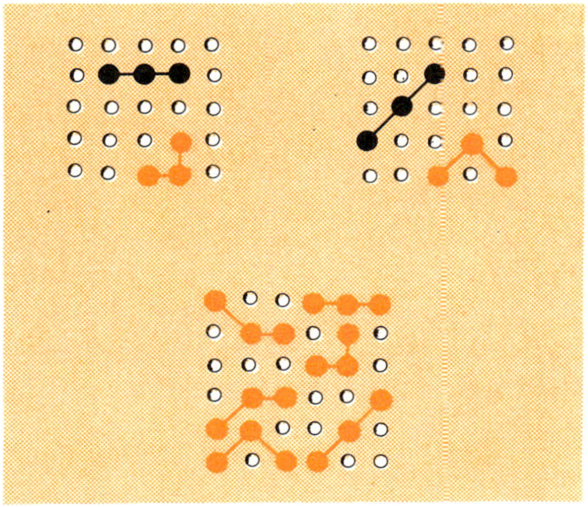

How many different patterns are there of 3 pegs where each peg is always next to at least one other?

We may decide to ignore rigid motions of the shapes in the plane of the board. We shall have to decide the meaning of 'next to'. The first diagram shows two patterns that arise from one decision. The other two diagrams show the two or six patterns that arise from other decisions.

How many different patterns will there be with 4 pegs in each case? 5 pegs? For each shape in the first case there is a corresponding shape in the second case, i.e. the one where 'next to' is always along a diagonal. These both correspond to polyominoes. (See p. 231.)

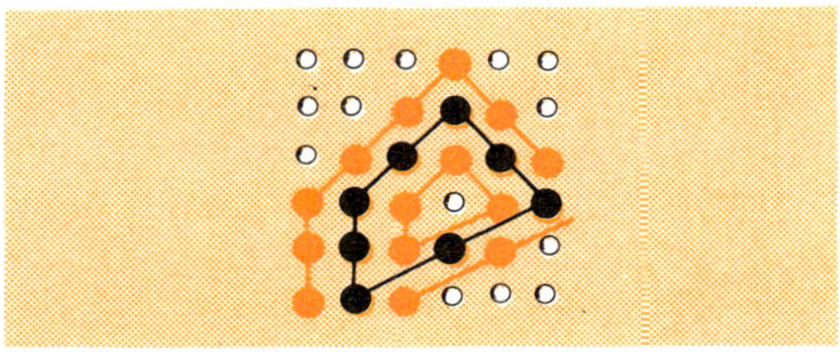

Make a shape with some pegs. How many pegs on the boundary? How many pegs fill in the holes that border this boundary? And the next lot of holes? Find the general form of the sequences of numbers generated in this way.

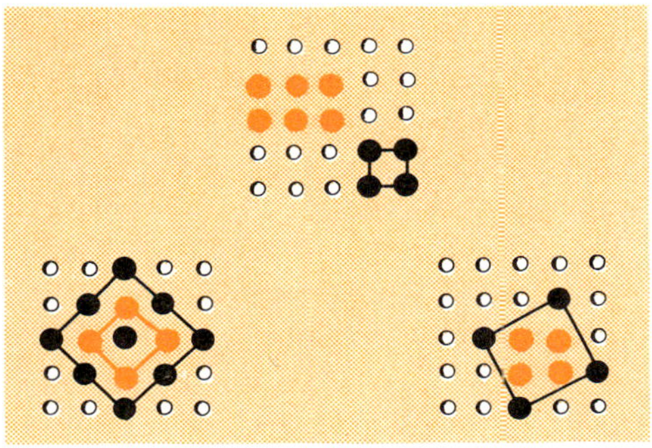

Make some rectangular shapes with pegs filling in all the holes inside or on the boundary of the rectangle. How many different rectangles can be made with a different number of pegs?

Where rectangles are restricted to those who lie in rows or columns of the board the situation raises questions about the composite or prime nature of numbers. It also generates the sequence of squares, 1, 4, 9, 16. . . . Note that if other rectangles are constructed the 'squares' produce other sequences, e.g. 1, 5, 13. . . .

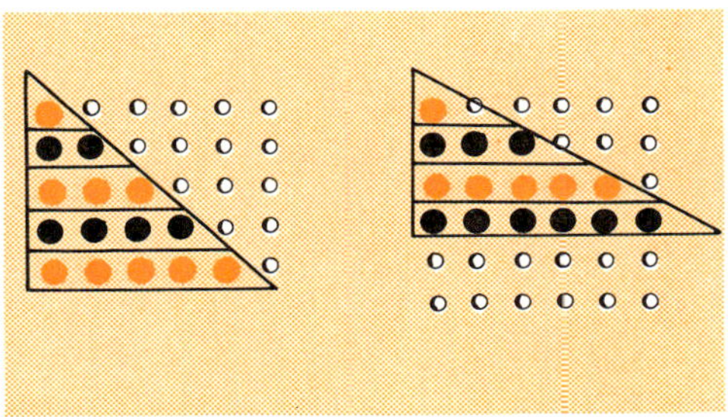

A triangular number of pegs is traditionally arranged as shown. The number of pegs in the successive triangles produces the sequence 1, 3, 6, 10. . . . The n-th term of this sequence will be the sum of the first n numbers.

In the last case the triangles are made by adding on rows with an extra peg each time. Consider 'elongated' triangles in which the rows have two extra pegs each time. What is the general term of the corresponding sequence?

Consider further elongations, e.g. having three extra pegs each time.

•	•	•	•	•	•
T_3	1	3	6	10	•
T_4	1	4	9	16	•
T_5	1	5	12	22	•
T_6	1	6	15	28	•
•	•	•	•	•	•

In general we may construct a sequence T_m of triangular numbers by adding m extra pegs in each row. The table of these sequences yields interesting patterns. The elongated triangular numbers turn out to be the square, pentagonal, hexagonal . . . numbers investigated by the Greeks. Various problems can be posed about these numbers, e.g. are there numbers that are triangular in two ways—say in T_3 and T_4.

Make up some problems.

A famous theorem asserts that every integer can be expressed as a sum of 3 triangular numbers, 4 square numbers, or in general $(m+2)$ numbers from the sequence T_m.

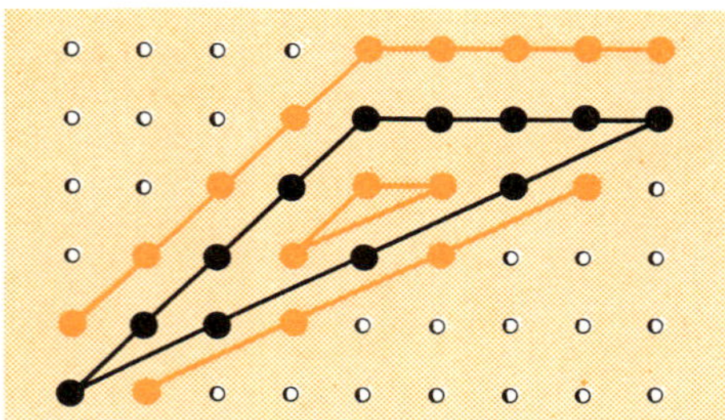

Suppose the investigation of triangular shapes had been set up in a more open way. Thus suppose three pegs are taken to form the vertices of a triangle. Insert pegs in all the holes inside or on the boundary of the triangle so defined.

How many pegs have you used altogether?

Generate a sequence of triangular numbers from similar shapes. Can you make a triangle with any number of pegs?

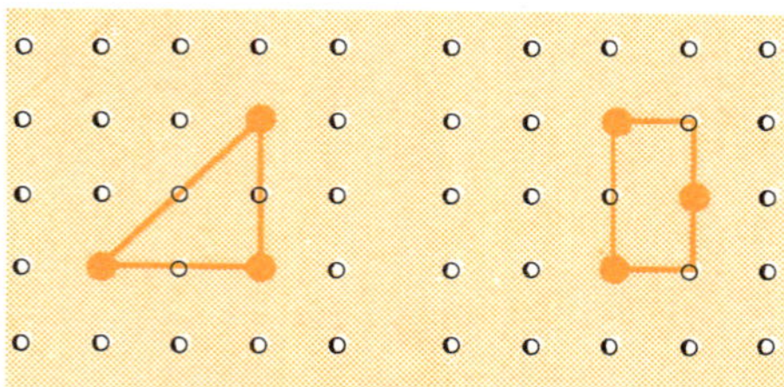

Can you place three pegs to form the vertices of an equilateral triangle? Any size of board is allowed. But your answer will depend on what you mean by the distance between two pegs. The diagram shows equilateral triangles with sides of 2 units for two different interpretations of distance. With the familiar interpretation of distance—the so-called Euclidean metric—no regular polygon other than a square can be constructed on the pegboard. Can you prove this?

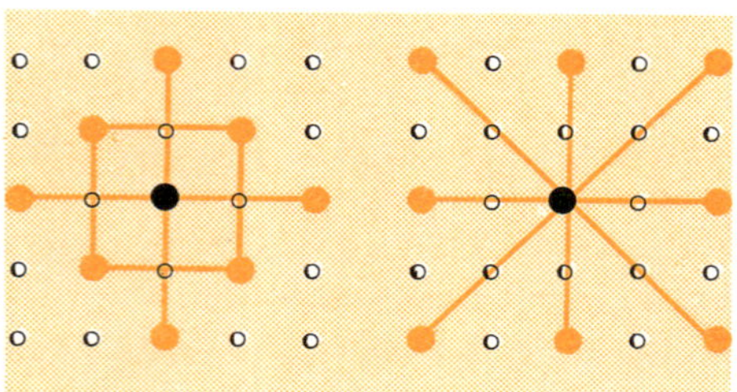

Place as many pegs as you can at a distance of 2 units from a given peg. The diagrams show two possible answers. The pegs inserted in each case form the *circle* of the peg board geometry concerned. . . .

See p. 209 for further investigations of some unfamiliar geometry.

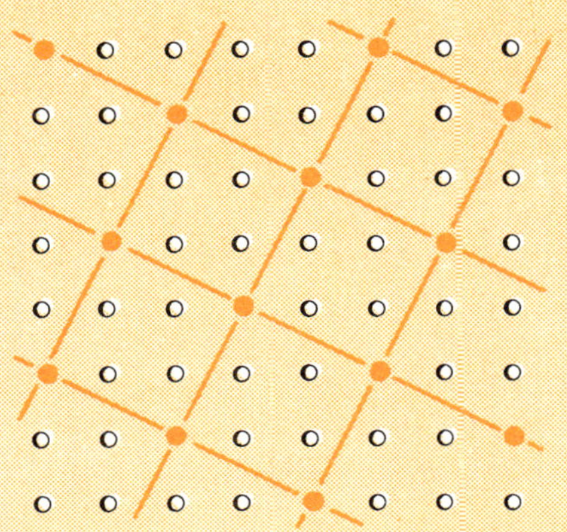

The pegs in the diagram make a square pattern superimposed on the pattern of holes. Each row has a peg in every 5th place but each row slips two places to the right from the row above.

Investigate in general the pattern obtained by placing pegs in every m-th place of a row, with a slip of n places from row to row.

For what values of m and n will the pegs lie in a square pattern?

This situation leads to some interesting arithmetical aspects of Pythagoras' theorem. (Cf. p. 208.)

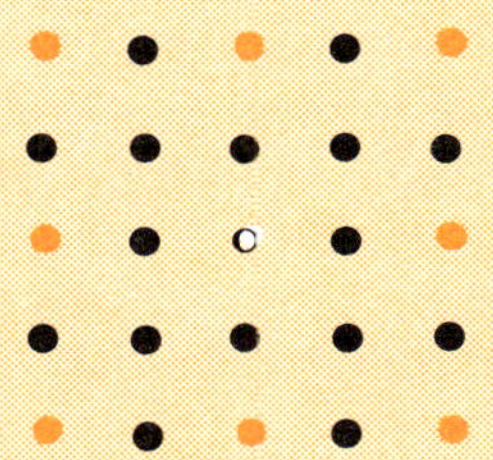

Study the pattern in the diagram. The dark pegs are *visible* from the centre hole, in the sense that there is no other peg directly between these pegs and the centre hole. The green pegs are in holes that are masked from the centre hole by a dark peg. Extend and investigate the pattern.

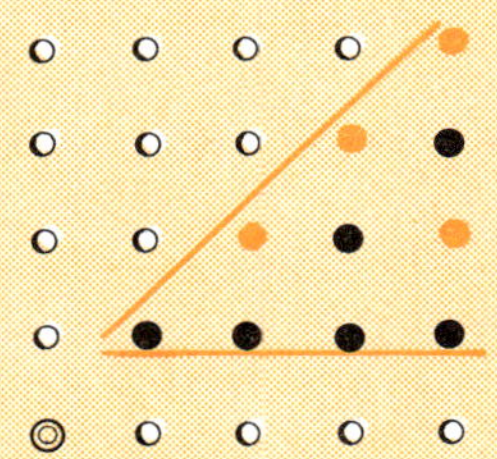

The pattern is symmetrical. The part in the half-quadrant indicated may be studied on its own. It is useful to consider the coordinates (m, n) of a dark peg with reference to the centre hole. It turns out that m and n are relatively prime, i.e. they have one common divisor only, namely 1.

Alternatively n/m is a fraction in its lowest terms.

An anti-clockwise sweep from the centre orders the fractions up to a particular denominator m in a series of increasing size, the so-called Farey series. The diagram indicates the series F_4. Such series have surprising properties, e.g. that $\frac{2}{3}$ and $\frac{1}{3}$ can be combined as $(1+2)/(3+3)$ to give the fraction $\frac{1}{2}$ which lies between them in the series.

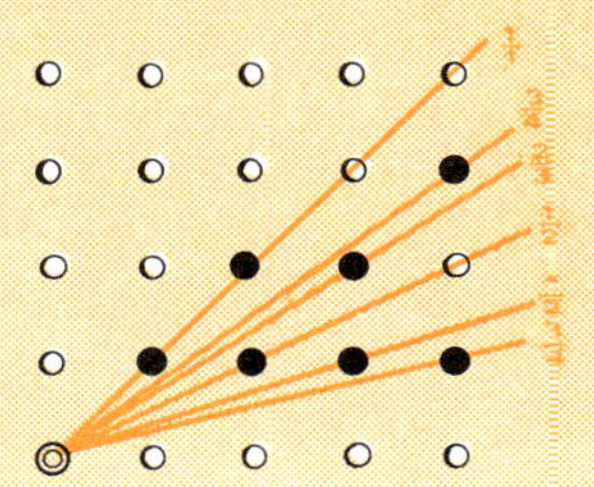

How many terms are there in the Farey series F_n?

We could perhaps answer this if we knew the number of pegs in each column of the half-quadrant array, i.e. the number of numbers relatively prime to m for each m.

Investigate the corresponding function. (Called Euler's function it crops up in many parts of number theory. See p. 103.)

Some interesting mathematical situations arise from various games and puzzles that can be explored on pegboard. See an A.T.M. pamphlet, *Pegboard games*, for a collection of suggestions.

playing cards

Playing cards can be obtained as cheaply as 15p a pack from most stationers or toy shops.

Blank playing cards which might be useful for various purposes can be obtained from Waddington (see p. 232).

Because of the many sub-sets, such as spades, red cards, aces, court cards contained in a set of cards, they are useful for probability experiments. From a well-shuffled pack, what is the probability of drawing an ace? A black card? A court card? Study the lengths of 'runs' of consecutive cards, such as 'runs' of the same suit.

Placing cards in specific arrays makes interesting investigations. For example, arrange the 16 cards consisting of the ace, 2, 3 and 4 of each suit in a 4×4 array, so that no two cards in the same column or row have the same value. In how many different ways can this be done?

Is it also the case in any of these arrangements that there are no two cards in the same row or column of the same suit?

Such arrays are called Latin squares; they have interesting practical applications as well as being related to group-tables.

Place two aces, two 2's and two 3's in such a way that there is 1 card between the aces, 2 cards between the 2's and 3 cards between the 3's. How many solutions are there? There are no solutions for 5 or 6 pairs but 25 solutions for 7 pairs.

Some card tricks can be mathematically analysed and are simple enough for pupils to investigate. One such trick is: ask someone to select a card from a pack of 21 cards, and return it to the pack after noting which card it is. Deal, in the normal way, into 3 piles with the cards face-up. The person who selected the card now indicates which pile it is in. Place this pile between the other two. Do this two more times. The selected card is now in the eleventh position from the top of the pack.

Trace the path of a selected card through this process. Why does the trick work?

Vary the number of cards in the pack. Does the trick still work? If not, can the number of deals, or the number of piles be adjusted to make it work?

Shuffling a set of cards gives various permutations of the cards. If the shuffling is done in a systematic way we can study these permutations and the number of repeated shuffles needed to restore the original order. For a set of n cards this number is the *order* of the permutation.

Some particular shuffles are:

(i) *Perfect shuffles:* Split a pack of cards into two equal piles, or as near equal piles as possible if there is an odd number of cards. A perfect shuffle interleaves the two packs exactly; this may be done by taking cards alternately from the bottoms of the two piles. There are some variations according to how one starts.

$$\begin{pmatrix} 1 & 2 & 3 & 4 & 5 & 6 \\ 1 & 4 & 2 & 5 & 3 & 6 \end{pmatrix}$$ cycles in 4

(ii) *Monge's shuffle:* Consider four cards a, b, c, and d in that order. b is placed on top of a; c underneath these two; d on top of these three. For more than four cards continue in the same way, placing alternately on the top and on the bottom of the pack.

$$\begin{pmatrix} 1 & 2 & 3 & 4 & 5 & 6 \\ 6 & 4 & 2 & 1 & 3 & 5 \end{pmatrix}$$ cycles in 6

(iii) *Invent your own shuffle.* One class invented the following: take the first card from the top of the pack. Place on top of this the card from the bottom of the pack, and on top of this the next card from the top and so on. There are variations of this particular shuffle.

$$\begin{pmatrix} 1 & 2 & 3 & 4 & 5 & 6 \\ 3 & 4 & 5 & 2 & 6 & 1 \end{pmatrix}$$ cycles in 6

(iv) *The cut:* Repeatedly cut the pack at the same place, e.g. between the 4th and 5th card. This is one of the simplest shuffles to analyse. The table shows the effect of a cut after the 4th card of a set of 6 cards. The original order is restored in 3 cuts.

$$\begin{pmatrix} 1 & 2 & 3 & 4 & 5 & 6 \\ 5 & 6 & 1 & 2 & 3 & 4 \end{pmatrix}$$ cycles in 3

The same cut for a set of 7 cards restores the original order in 7 cuts. What is the general rule for this and other cuts?

Arrow graphs can be a useful way of recording the effect of a shuffle. (See p. 164.) The graphs raise various further problems, e.g. combining two graphs to get a third; finding a formula for the number of shuffles required to restore original order; other ways of shuffling; combinations of shuffles, and so on.

Place an ace, two, three and four in a circle in any order. Begin counting at any card and count round the circle. Stop the count at any stage where the number of the count appears on the card being counted, i.e. if the m-th card from the start has the number m on it. In this case take up this card and begin counting again at the next remaining card.

There are 5 initial arrangements of the four cards which lead to the removal of all four of the cards. What are they? How many arrangements lead to the removal of only three cards? No cards?

Investigate what happens when the game is played with five cards—from an ace to a 5. Try other sets of cards, including ones that are not consecutive from an ace. Cf. p. 80.

Take two packs of cards and compare them card by card. How likely do you think it is that there will be no stage at which the same card is turned up from each pack?

It is perhaps surprising that it is in fact more likely than not that there will be a coincidence. The situation is related to that of finding the number of ways of placing objects in the wrong pigeon-holes. This derangement problem is analysed in various standard texts. The theory shows that in the case of a large number of comparisons of cards the probability that no coincidence occurs is approximately $1/e$, where e is the exponential constant 2·718. . . . The probability is thus less than 0·5.

This result could be used to set up an experimental determination of the value of e. Shuffle and compare the two packs a 100 times. If N is the number of occasions when no coincidence occurs then $100/N$ gives an approximate value for e. A closer approximation might be expected if the packs are compared a thousand times. This is an example of the so-called Monte-Carlo method, a particular useful way of attacking certain mathematical problems with electronic computers which can repeat a counting operation many thousands of times.

stationery

This section is mainly concerned with items for general use in the mathematics room though some have been included because of specific uses mentioned elsewhere in this book. Most of the items mentioned are available in local stationery shops but it is worth consulting catalogues from general educational suppliers. Such firms as Arnold, Brightway, E.S.A., Galt or Philip & Tacey—for addresses see p. 232—produce comprehensive catalogues each year and will send them free of charge to teachers on request. Another useful source is the L.E.A. Supplies department which usually has a special list for schools.

adhesives Some adhesives are cheaper if bought in bulk but for classroom use they need to store well when transferred to smaller containers. For some purposes, such as model making from card, it is essential to use quick-drying adhesive.

Adhesive tape (e.g. sellotape) is most economically used if bought in large rolls and stored in a dispenser. Double sided adhesive tape is also available.

For card and paper models: Copydex, which can be purchased in small tubes or containers holding up to a gallon, or balsa cement, sold in small tubes.

For wood: any contact adhesive such as Evostik, Bostik or Gloy Multi-glue.

For mounting: Cow-gum is a special rubber solution; any surplus can be removed without leaving a mark and paper stuck to glass windows can be peeled off easily.

Gloy is a good glue for general use.

boxes Sturdy cardboard boxes, which are sold in various sizes, are almost essential for storing equipment. Some catalogues list them only in the Infant School section. Cheaper, but less durable, are shoeboxes. These can often be collected free of charge from local shops.

card Sold in various sizes, thicknesses and colours. It is more economical to buy large sheets but these raise storage problems as it is essential that the card is stored flat. For model making 6-sheet card is recommended, for linkages it is necessary to have thicker card, say 12-sheet.

clips It is always useful to have a supply of ordinary paper clips. For more permanent fastening, say a class booklet or a student's project, a stapler is most suitable. One sturdy model with a long arm is a better investment than several small ones.

A stapling gun provides a quick, neat way of mounting display material, but the staples are a nuisance to remove from the notice board. Drawing pins can be unsightly; alternatives are ordinary needlework pins—which are very cheap—or mapping pins.

cutting tools The cheaper rounded scissors that are widely available have limited use in the mathematics room but larger ones, say 18 cm, can be used for cutting card of 6-sheet thickness as well as for general work. A portable block or rack for housing scissors makes distribution and collection easier. To cut thicker card, a craft knife and metal safety rule are required. Squares of hardboard make suitable cutting boards. Guillotines to take larger sheets of paper or card are expensive but it may be worth having one with a cut of about 30 cm.

duplicating materials The teacher who does not rely on a series of textbooks soon realizes the need for stencilled material. One of the simplest forms of duplicating is the Banda spirit machine and the mathematics department might consider having one of its own. With ink duplicators, such as Roneo and Gestetner, it is possible to obtain further copies from a stencil at a later date if it is kept in good condition in a stencil cabinet. Sophisticated ways of preparing stencils from printed material are now available.

elastic bands Required for geoboard work and usually stored on the boards. Bands are often sold in packets of mixed sizes and colours but packets of the same size and colour (e.g. Pym bands) can be purchased from some stationers.

Shirring elastic which stretches easily and is useful for work with geostrips can be bought from haberdashery stores.

folders Sometimes students' work might take the form of a booklet and a useful way of making one is to staple the work into a manilla folder. If the work is to remain in loose sheets a pocket folder is more suitable, though more expensive.

gummed paper It is much cheaper to buy coloured gummed paper in packets of 100 from the educational suppliers. Various shapes and sizes, including polygons, circles and strips, are available from Charles Ltd (cf. p. 174).

paper A variety of type and size of paper will give students an opportunity to record and present their work in different ways. The local paper mill is often a cheap source for large quantities.

Cartridge paper: obtained in various sizes, including large sheets for display work. It is useful for paper-folding activities and for model making.

Graph paper: paper printed with grids of various sizes is an important basic tool for mathematical activity (see p. 206). Paper ruled in 1 cm or 2 cm squares will be required in quantity. Special graph paper such as isometric, polar and logarithmic paper may be obtained from Chartwell (p. 232).

Tracing paper: best obtained in packets rather than rolls as it usually needs to be kept flat. Tracing paper ruled in squares can be obtained from Philip & Tacey (p. 232).

pencils and pens There are many occasions when work can be made more attractive by the use of felt, or fibre tip, coloured pens. A block of wood in which holes have been drilled to hold pencils or pens is a useful way of checking numbers at the end of a lesson. Chinagraph pencils are used for writing on highly polished surfaces such as acetate sheet; marks can be removed with tissue paper. The lead of these pencils breaks easily and they should be sharpened with a craft knife.

paints Acrylic paint is recommended for painting models. It is sold in tubes (e.g. Rowney PVA or Rowney polymer); can be used neat or thinned with water, and is quick drying.

tiles

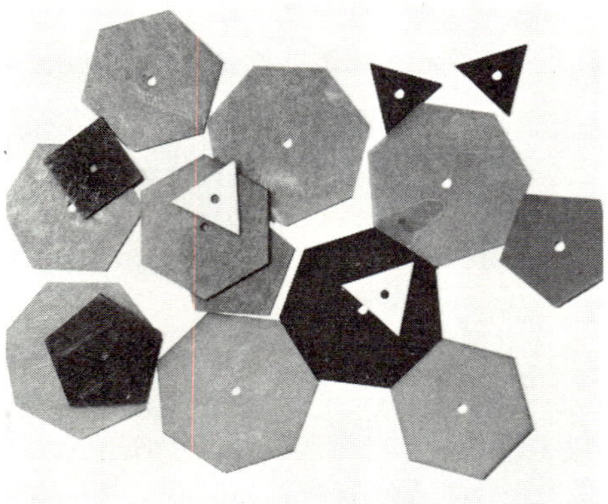

A good supply of tiles is essential. Most of the sets on the market have regular shapes only—it is worth considering others. Regular polygon tiles should be compatible, i.e. built up on a common length of side so that they fit together when laid edge-to-edge. Cardboard tiles are produced by E.S.A. They can be made in sufficient quantity from templates. Coloured gummed paper shapes are attractive and inexpensively produced in various sizes by Charles. There are various sets of wooden tiles available. One of the best is the set of *Pattern Tiles* from the Elementary Science Study (U.S.A.), distributed in U.K. by McGraw Hill. Tiles in plastic are produced by Arnolds and other firms. It is worth considering having a set made to one's own specifications by a small plastics firm. We have found the following set to be particularly useful. Made from virtually unbreakable plastic in various colours the set consists of 24 equilateral triangles, 24 squares and 12 each of the regular polygons with 5, 6, 7 and 8 sides. All sides are 5 cm. This set was made by Abbey Products at £3·50. (For addresses of firms mentioned, see p. 232.)

Make some patterns on the floor or table.

Rules may be formulated for laying the tiles. Thus tiles may be laid edge-to-edge, corner to corner, half-edge to half-edge and so on.

In some cases tiles can be laid to cover the plane completely leaving no spaces. In other cases an arrangement of tiles leaves spaces which it may or may not be possible to fill with other available tiles.

An arrangement of tiles which leaves no spaces we shall call a *tiling*. (Other names in use are 'tessellation' and 'mosaic; sometimes these are restricted to tilings with regular polygon tiles. The reader should bear in mind irregular alternatives.)

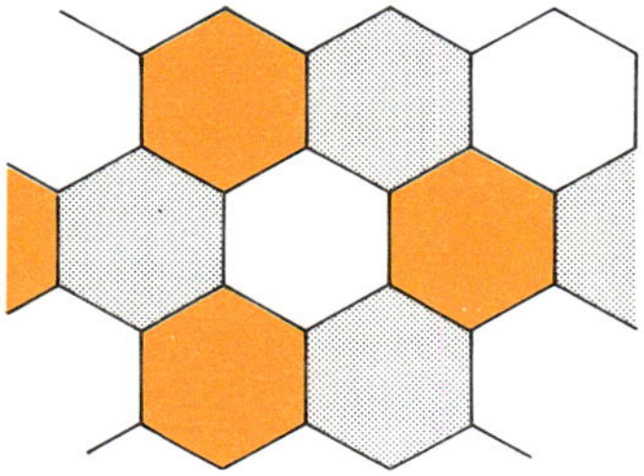

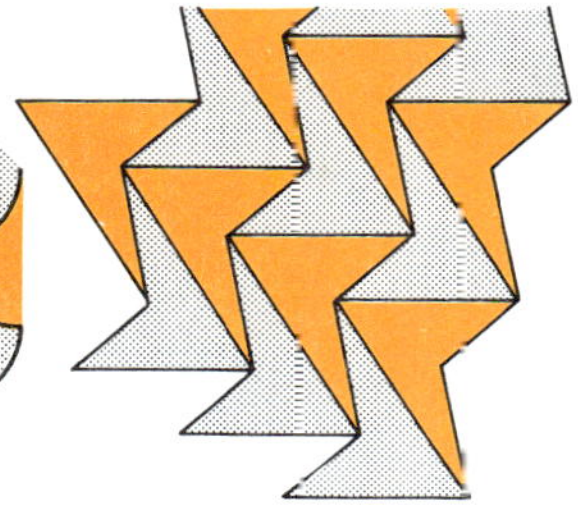
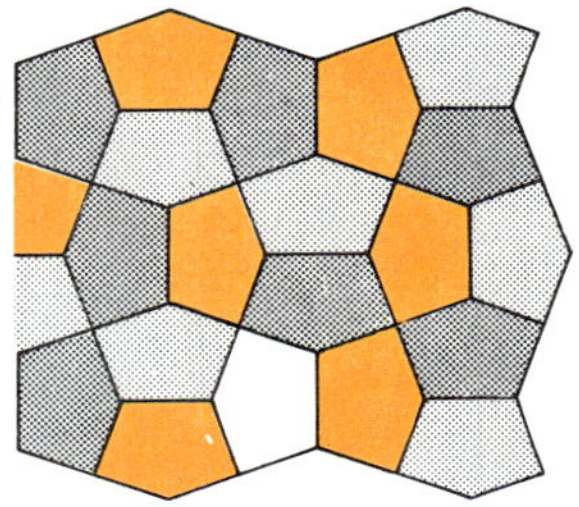

Find as many tilings as you can with only one shape of tile.

Investigate the different tilings which can be made with a 60° set square; with a 45° set square.

Will any quadrilateral tile?

A regular pentagon will not tile on its own. Can you find a pentagon that will?

What sort of hexagons will tile?

When laid edge-to-edge some regular polygon tiles fit completely leaving no spaces; others leave spaces which cannot be filled by other regular tiles. When laid corner-to-corner some regular polygon tiles fit together leaving regular star-shaped spaces. Thus the hexagon with its corresponding star polygon will form a tiling. Which other shapes will do this?

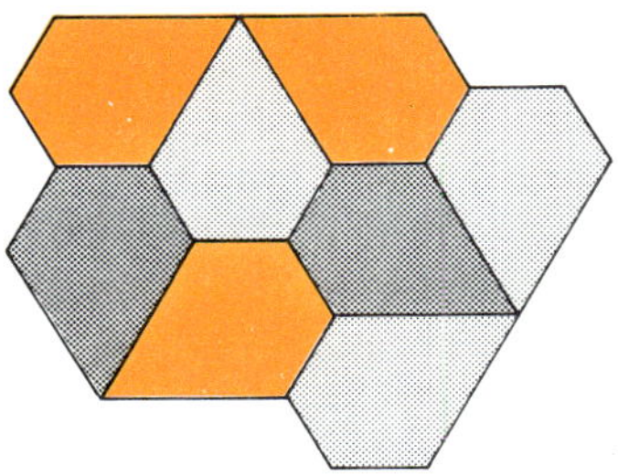
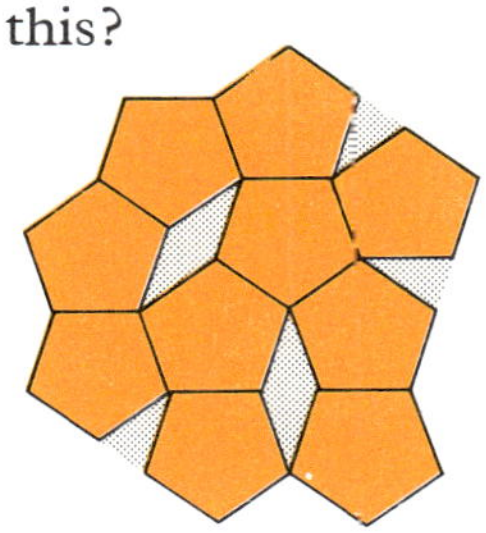
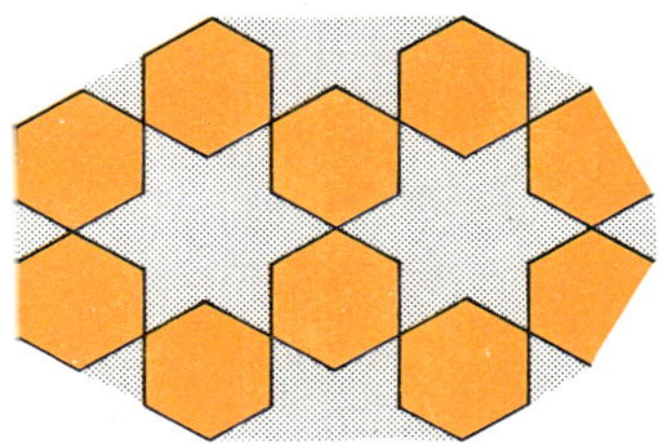

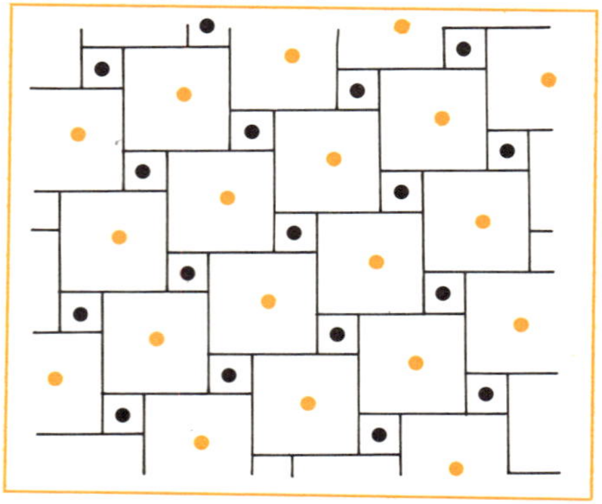

When regular polygons are laid with edges partly touching the spaces produced may still be regular polygons. Thus squares may be arranged so as to leave spaces which may be filled by another size of square.

Can other shapes be tiled in this way?

Investigate the tiling of squares further. In particular consider the super-imposed tiling implied by the pattern formed by the centres of all the squares. A vivid dissection proof of Pythagoras' theorem is available.

Clearly there is an infinite number of tilings when no limitations are imposed on the shape or arrangement of the tiles. Imposing some restrictions produces some interesting problems.

Consider polygon tilings only.

It seems natural to demand at least that the tiling should have a repeating pattern. There are still unlimited possibilities, though whatever shapes are used the repeating pattern can in fact be classified by its symmetries into one of 17 types—the so-called wallpaper groups.

A repeating pattern means that there will be some restriction on the number of different shapes of tile and also on the number of different arrangements of shapes at each corner or vertex.

A tiling with m different shapes arranged so that there are n different arrangements at vertices will be called an (m, n)-tiling. Corresponding to every such tiling there will be a dual (n, m)-tiling with n different shapes and m different vertices.

How many (1, 1)-tilings can you find?

In one sense there are only three, the triangle, quadrilateral and hexagon tilings, though these can be distorted into various forms. The triangle and hexagon tilings are dual, and the quadrilateral tiling is self-dual.

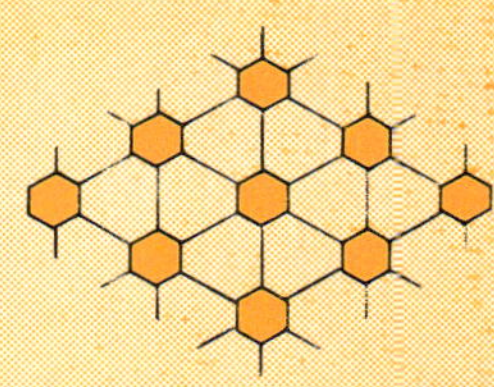

With $n=1$, i.e. one type of vertex only, a tiling may be represented symmetrically by regular polygons. There are 11 such tilings—the so-called regular tessellations.

These are shown on page 211.

Take any one of these tilings. If the centres of any two adjoining edges are joined, these joins produce a *dual* tiling. Find the 11 dual tilings. These have $m=1$, i.e. only one shape is used in the tiling.

How many tilings can you find using regular polygon tiles?

There are 17 possible dispositions of regular polygons at a vertex but some of them cannot be repeated to form a tiling. Why not?

Besides the 11 (m, 1)-tilings there are a further 15 'semi-regular' tilings with n=2 or 3, i.e. having two or three different types of vertices.

Tilings can be coloured in various ways to form fascinating designs.

Consider a square tile divided diagonally into two colours—black and white are effective. What patterns can be produced from these tiles? What would be a suitable definition of a 'regular' pattern? How many such regular patterns can be formed?

Having composed a pattern what transformations are required to change it into another by interchanging tiles, interchanging rows or columns, rotating tiles and so on?

Take 36 of these tiles and arrange them in a 6×6 square. Each tile is labelled by a coordinate pair with reference to, say, the bottom left one which is labelled (1, 1). Generate random number-pairs by successive double throws of a die (or by throwing two differently coloured dice). Thus if the die shows a 5 on the first throw and a 2 on the second this yields the pair (5, 2). The square in this position is now rotated clockwise through one right angle. The process is repeated.

Study the transformation in the pattern over a lengthy sequence of throws.

Would you expect a repeat?

Devise other ways of producing random sequences of patterns.

In how many ways can a square, divided by both its diagonals, be coloured using 2 colours? 3 colours?

What patterns can be produced using some or all of these tiles?

Consider some other colourings of square tiles.

Consider other shapes of tile (cf. the work on dominoes, p. 188).

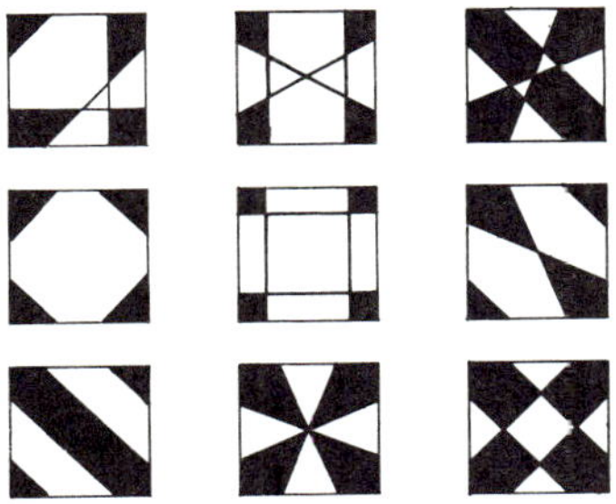

An interesting set of 9 tiles is produced by Invicta Plastics (see p. 232). These are marketed under the name of Pollypattens; they raise some interesting questions:

Why are there only 9? What are the rules for making them? What patterns can be made using only one type? How does the symmetry of the tile affect this number? What patterns can be made if the tiles are joined half-edge to half-edge? What would triangular pollypattens be like?

Design your own set of tiles.

A further activity which might be described as tiling is that of fitting polyominoes together. Polyominoes are the shapes which can be made by joining a given number of squares edge-to-edge. Thus consider 5 squares. How many different shapes (pentominoes) can be made when they are joined in this way?

There are in fact 12 pentominoes and these can be fitted together to form a 10×6 rectangle. This is used as a plastic puzzle set marketed by Pan Playthings.

Could other rectangles be made?

Which of the pentominoes, or combinations of pentominoes could be used for a tiling?

Investigate other tiling problems with various polyomino shapes.

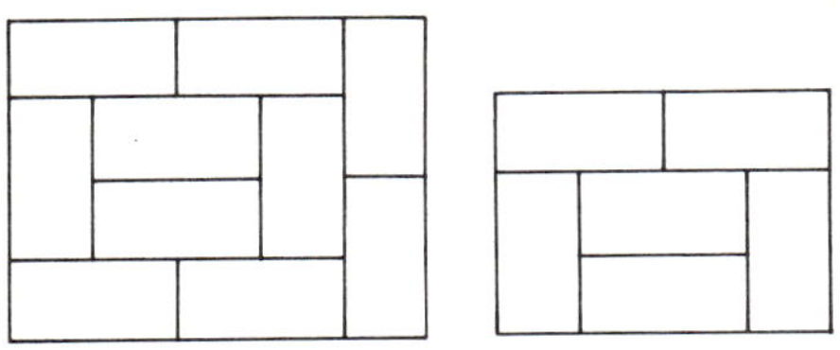

The simple rectangular polyomino will tile in various different ways. A convenient way of investigating these is to use Cuisenaire rods. What tilings are possible with the red rods? The green rods? (For further reference to polyominoes see p. 118).

some addresses

Abbey products, Chudleigh, Nr. Exeter, Devon.
Addo Ltd., (calculators), 5 Worship St., London E.C.2.
Altair designs, 52 Borough High St., London S.E.1.
Arnold, E. J., & Son Ltd., (ed. supplier), Butterley St., Leeds 10.
A.T.M., Market St. Chambers, Nelson, Lancs.
A.T.M. Filmloops, 64a Wharf Rd., Grantham, Lincs.
Blackwell's bookshop, Broad St., Oxford.
Block & Anderson Ltd., (calculators), Cambridge Grove, London W.6.
Bradley, R., 62 West Farm Av., Ashtead, Surrey.
Brown & Sons, Ltd., Perth St. West, Hull.
Burdett, N., (ed. supplier), 44 Main St., Kilby, Leics.
Brightway Aids, (ed. supplier), 123 Pollard St., Manchester.
British Film Institute, 81 Dean St., London W.1.
Charles & Son Ltd., 1 Woodbridge St., Clerkenwell Green, London E.C.1.
Copyprint on Education, 87 Borough High St., London S.E.1.
Cuisenaire Co. Ltd., (ed. supplier), 40 Silver St., Reading.
Dillon's bookshop, Malet St., London.
Electronix Co., 171–5 Southampton Way, London S.E.5.
Elford Plastics Ltd., Wood St., Elland, Yorks.
E.S.A. (Educational Supply Association), Pinnacles, Harlow, Essex.
Fisher, D., Ltd., Thorp Arch Trading Estate, Boston Spa, Yorks.
Galt, J., & Co. Ltd., Brookfield Rd., Cheadle, Cheshire.
Gateway Films, 470 Green Lanes, London N.13.
Heffer's bookshop, Petty Cury, Cambridge.
Ian's Ltd., Ferguson's Lane, Benwell Village, Newcastle-on-Tyne 5.
I.B.M., 389 Chiswick High Rd., London W.4.
Invicta Plastics Ltd., London Rd., Oadby, Leicester.
McGraw Hill Co. Ltd., Shoppenhangers Lane, Maidenhead, Berks.
Morton Products Ltd., Lady's Lane, Northampton.
Muldivo Calculating Machines Co. Ltd., Salisbury Square, London E.C.4.
National Computing Centre, Quay St., Manchester 3.
National Film Board of Canada, 1 Grosvenor Square, London W.1.
Pan Playthings Ltd., 4 Rodney St., London N.1.

Philip & Tacey Ltd., (ed. supplier), 69/79 Fulham High St., London S.W.6.
Proops Bros. Ltd., 52 Tottenham Court Rd., London W.1.
Rank Film Library, 1 Aintree Rd., Perivale, Greenford, Middlesex.
R.J.M. Exports Ltd., Glebe House, Fossebridge, Glos.
Spear, J. W., Ltd., Enfield, Middlesex.
Taskmaster Ltd., (ed. supplier), 105/7 Clarendon Park Rd., Leicester.
Thin's bookshop, South St., Edinburgh.
Tor Instruments Co., 4 Steps Lane, Torquay, Devon.
Trendon Ltd., Old Malton, Malton, Yorks.
Waddington, J., Ltd., Berkeley Square, London W.1.

IV Appendix

IV Appendix

on mathematical education

Mary Boole was the wife of the mathematician George Boole, who played such an important part in the development of modern algebra. In the twenty years after his death Mary Boole supported and raised a family and read, corresponded, discussed and wrote, over a wide range of themes. Her collected works amount to more than fifteen hundred pages. Among them is found a regular and passionate insistence on the real—but widely ignored—significance of her husband's work, and a vision of mathematical education that is powerfully relevant, but still unrealised, today.

We give a few extracts from her writings in this appendix as a final postscript. The quoted passages end with a page number reference to the collected works mentioned in the book list on p. 64.

'If the child uses the nominative where he should use the accusative, and is not at once corrected, that is so much to the bad for his future progress; if he can be got not to remember a time when he used the word wrongly, that is so much to the good. But in science there are, there can be, no absolutely right impressions; our minds are big enough to grasp any natural fact as a whole; everything depends upon drawing right conclusions from combinations of impressions, each of which is in itself inadequate and partially misleading; and if the pupil is to be got into scientific methods, that is what he must be trained to do. And in order that he may learn to do it, it is sometimes necessary that each of a succession of "wrong" impressions should have time to register itself on the brain and become part of its available stock. Such a statement may naturally convey to the scholastically mind trained in classical traditions an impression of disorderliness, but it does not imply disorder. Up-to-dateness is the cause of disorder; the haste, the greed to efface rapidly each partial impression, when we have nothing to substitute for it but some other impression equally partial, is not only unscientific but eminently disorderly.' (882)

on influence

'The cultivation of the mathematical imagination depends chiefly on the child being put into the right attitude towards mathematical conceptions in his earliest years; and, after that, on the right use being made of certain nodes or critical points which occur here and there in each branch of mathematics, and which should be dealt with in a quite different manner from the rest of the course. These form the revelation crises of the pupil's mathematical history; when he draws near one of these the human teacher should carefully withdraw his influence, and simply watch to see that no seriously false impression is being formed. His object should be to efface himself, his books, and his systems; to draw aside a curtain from between the child and the process of discovery, and to leave the young soul alone with pure truth.' (919)

'In all ordinary works of education, we may notice constant discussion about the particular mode in which influence should be brought to bear, whether by individual commands, unvarying rules, hopes or fears about consequences, appeals to affection, the contagion of example, or that more subtle form of influence called by the pious intercessory prayer, and by modern science suggestion or telepathy. The whole discussion usually turns on the rival merits of the various modes of bringing influence from without to bear upon the pupil. It seems assumed that it is always legitimate to exert influence. In mathematics, however, the main question kept in view is: When may the teacher exert influence?

For mathematical purposes, all influences from without, which induces the pupil to admit a principle as valid before his own unbiassed reason recognizes its truth, come under the same condemnation.' (727)

on teacher lusts

'The teacher (whether school-teacher, minister of religion, political leader, or head of a family) has a desire to make those under him conform themselves to his ideals. Nations could not be built up, nor children preserved from ruin, if some such desire did not exist and exert itself in some degree. But it has its gamut of lusts, very similar to those run down by the other faculties. First the teacher wants to regulate the actions, conduct, and thoughts of other people in a way that does no obvious harm but is quite in excess both of normal rights and of practical necessity. Next, he wants to proselytise, convince, control, to arrest the spontaneous action of other minds, to an extent which ultimately defeats its own ends by making the pupils too feeble and automatic to carry on his teaching into the future with any vigour. Lastly, he acquires a sheer automatic lust for telling other people "to don't", for arresting spontaneous action in others in a way that destroys their power even to learn at the time what he is trying to teach them. What is wanted is that we should pull these three series tight so as to see their parallelism, and not go on fogging ourselves with any such foolish notion as that sex-passion is a lust of the flesh and teacher-lust a thing in itself pure and good, which may legitimately be indulged in to the uttermost.

Few teachers now are so conceited as not to know that they have a great deal to learn, and that their methods need revising and improving, but the majority are seeking for improved methods of doing more of what they are already doing a great deal too much of. The improvement which they most need is to be brought under conviction, to be made to see their conduct, their aims, their whole attitude towards their pupils and their work, in the light reflected on them from those of the drunkard and the debauchee.' (1412)

Postscript

Presumably you who are reading these words are involved in the teaching of mathematics.

Why?

There must be many possible answers to the question but whatever these are, please consider them carefully. It is difficult to consider what you are teaching, whom you are teaching, and how you are teaching, without paying some attention to the question of why you are teaching.

One day a friend entrusted a jar to the Hodja and asked him to keep it for him until he came back. A few days passed, and the Hodja, who had been very curious to know what was in the jar, grew more and more impatient. Finally he broke off the lid, and looked inside. The jar was full of honey. The Hodja dipped his fingers in, and tasted the honey. It was excellent. The Hodja replaced the lid, and went about his business. It was not long, however, before his mind returned to the delicious honey. Off came the lid, in went the finger, on went the lid, and the Hodja went about his business. So it continued until the jar was wiped clean of any trace of honey. The owner returned and asked for the jar he had entrusted to the Hodja. It felt rather light and he looked inside.

'Hodja,' he exclaimed, 'Where is my honey?'

'How nice it would be if you hadn't asked that question,' sighed the Hodja, 'and I didn't have to reply.'

Index

Notes